EMERGING SYNTHESES
IN SCIENCE

EMERGING SYNTHESES IN SCIENCE

PROCEEDINGS OF THE FOUNDING WORKSHOPS
OF THE SANTA FE INSTITUTE
SANTA FE, NEW MEXICO

Volume I

David Pines, *Editor*

Department of Physics, Loomis Laboratory
University of Illinois, Urbana

Advanced Book Program

CRC Press
Taylor & Francis Group
Boca Raton London New York

CRC Press is an imprint of the
Taylor & Francis Group, an **informa** business

First published 1988 by Perseus Books Publishing

Published 2019 by CRC Press
Taylor & Francis Group
6000 Broken Sound Parkway NW, Suite 300
Boca Raton, FL 33487-2742

First issued in hardback 2019

CRC Press is an imprint of the Taylor & Francis Group, an informa business

Visit the Taylor & Francis Web site at
http://www.taylorandfrancis.com

and the CRC Press Web site at
http://www.crcpress.com

Library of Congress Cataloging-in-Publication Data

Emerging syntheses in science.

    1. Science—Congresses. 2. Santa Fe Institute
(Santa Fe, N.M.)—Congresses. I. Pines, David,
1924– . II. Santa Fe Institute (Santa Fe, N.M.)
Q101.E49   1987   500   87-24121
ISBN 0-201-15677-6

ISBN 13: 978-0-367-32045-4 (hbk)
ISBN 13: 978-0-201-15686-7 (pbk)

This volume was typeset using TEXtures on a Macintosh computer.

DAVID PINES
Urbana, IL, September, 1987

Foreword

The Santa Fe Institute, as a key element in its founding activities, sponsored two workshops on "Emerging Syntheses in Science," which took place on October 5–6 and November 10–11, 1984. Each workshop began with a description by Murray Gell-Mann of the concept of the Institute. Subsequent speakers described aspects of emerging syntheses which might prove relevant to the future development of the Institute and George Cowan described some of the initial steps which are being taken to create the Institute. In the course of the talks, and discussions which ensued, a number of possible future directions for the Institute were explored. Networks, which might tie together researchers in a newly emerging synthesis, using both traditional and innovative forms of communication, ranging from workshops and the exchange of graduate students and postdoctoral fellows to computer links, emerged as one of the initial foci of the Institute activity.

There was unanimous agreement among the participants that Professor Gell-Mann's keynote address and the ensuing talks were of such high quality and general interest that it would be highly desirable to publish these for broader distribution. I agreed to serve as Editor of the Proceedings and chose an informal format for this volume as a way of making the proceedings rapidly available at modest cost. To convey the character of the workshops, contributors were encouraged, in writing up their talks, to follow the same kind of informal approach which characterized their presentations. The grouping of the talks is intended to reflect some of the many connections between apparently different problems which became evident during the workshops, while the flow chart below illustrates some further connections. For those speakers who were not able to contribute a manuscript, brief summaries of their remarks are appended.

It gives me pleasure, on behalf of the Board of Trustees of the Santa Fe Institute, to thank the contributors to this volume for their rapid response to my pleas for manuscripts, and to thank the Carnegie Corporation of New York and the John D. and Catherine MacArthur Foundation for their financial support, which has made possible both the workshops and the publication of this volume. Special thanks go to Karie Friedman and Ronda Butler-Villa for their assistance in editing this volume, and to Françoise Ulam for her translation of the article by Dr. Schützenberger.

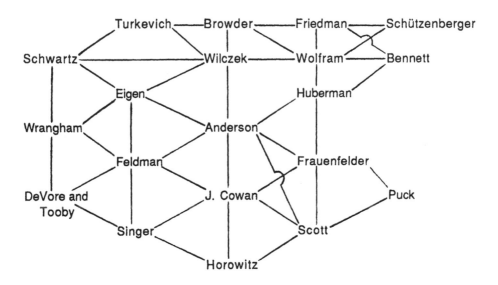

FIGURE 1 FLOW CHART, illustrating some of the connections between talks and topics discussed at Santa Fe Institute Workshops on "Emerging Syntheses in Science."

Contents

EMERGING SYNTHESES
IN SCIENCE

MURRAY GELL-MANN
California Institute of Technology

The Concept of the Institute[1]

It is a pleasure to welcome so many old friends and a few new ones to this beautiful place, kindly lent us by Douglas Schwartz of the School of American Research. We would like to hear your reactions to the proposal we are making for setting up the Santa Fe Institute and to hear your ideas about how to structure it, what kind of intellectual problems it should address, what kinds of arrangements should be made for its governance, and what should be the first steps in establishing it.

It is usually said that ours is an age of specialization, and that is true. But there is a striking phenomenon of convergence in science and scholarship that has been taking place, especially in the forty years since the Second World War, and at an accelerated pace during the last decade. New subjects, highly interdisciplinary in traditional terms, are emerging and represent in many cases the frontier of research. These interdisciplinary subjects do not link together the whole of one traditional discipline with another; particular subfields are joined together to make a new subject. The pattern is a varied one and constantly changing.

In order to discuss a few examples of diverse character, I shall start from subjects close to my own and then move further away. I hope you will forgive me

---

[1] Talk given at the Founding Workshops of the Santa Fe Institute, held at the School of American Research, Santa Fe, New Mexico, in November 1984. Revised version.

for talking about matters far from my area of expertise and will correct whatever howlers I make in the course of doing so. Also, I apologize for mentioning in this introduction, for lack of time and space, only some of the emerging syntheses about which we shall hear and only some of the distinguished speakers who will discuss them.

Elementary particle physics and the cosmology of the early universe are the twin pillars on which all the laws of natural science are, in principle, based. These two fundamental subjects have practically merged in the last few years, especially on the theoretical side. In the earliest fraction of a second in the history of the universe, if we look at time running backwards, we go from an easily comprehensible quark soup (a few moments after the beginning) to an earlier era in which the conditions are so extreme that, if we could observe them, they would test our speculative ideas about unifying all the physical forces including gravitation. I should mention that in the last few weeks these ideas of unification have become much more specific. The hope, the very bold speculation, that we might actually find a general theory of all the elementary particles and forces of Nature is encouraged by recent developments in superstring theory.

Many of the mysteries of the universe seem to be tied up with particle physics.

As Frank Wilczek will probably tell us, the mystery of the smallness or vanishing of the cosmological constant, which is the value of the energy density of the vacuum, is intimately connected with particle physics. The mystery of the dark matter in the universe, which must outweigh visible matter by at least a factor of ten, is now believed to be in the domain of particle physics, since much of the dark matter may consist of hypothetical new particles such as photinos or axions

Meanwhile, the trend toward divorce between physics and the frontier of pure mathematics, which went on for decades after the end of the nineteenth century, has been reversed. The description of elementary particle interactions and the attempts to unify them connect with the central part of pure mathematics, where algebra, analysis, and geometry come together, as in the theories of fiber bundles, of Kac-Moody algebras, and so forth, Frank Wilczek may address that topic, too,[2] and Felix Browder will probably touch on it, as he discusses important parts of mathematics that are applicable to science.

The other examples will be drawn from the study of highly complicated systems. First of all, in the life sciences, a transformation has taken place in recent years that has been so dramatic as to impress itself on everyone, scientists and the general public. Some central themes in biology and medicine will be addressed by Ted Puck. Of course, a discussion of the revolution produced by advances in molecular biology needs no introduction, but I should like to quote some remarks made last spring by a distinguished worker in that field A few months ago, the National Academy of Sciences gave a party to celebrate the winning by American citizens of several Nobel Prizes, as well as the Swedish Riksbank Prize in Economic Science awarded in memory of Alfred Nobel. I was invited to speak, along with three other members of the Academy, and I chose the same subject as today's. In fact, at least three

---

[2] In fact, Wilczek's paper treats a different but related interdisciplinary subject.

of the four speakers that afternoon had chosen independently to address related subjects. David Baltimore, who preceded me, and Herb Simon, who followed me, both discussed the remarkable trends in science with which we are concerned here, and I shall take the liberty of quoting from David Baltimore's remarks

"The first place to start is to look at what's happened to biology in the last ten years. About ten years ago, the field went through a watershed, because up to then, the precise tools available for dissecting genetic structure and understanding biological organization were really only applicable to microorganisms, which provided useful model systems but couldn't answer the pressing questions of the organization of systems in human beings and other mammals

"Then, about that time, a variety of new techniques were developed that allowed us to get at the molecular details of higher systems, and overnight what had been seen as impossible became eminently feasible, and the methods which we generally call the recombinant DNA methods changed our whole perspective on what we could think about and do, and that has had many consequences, one of which has been the focus of molecular biology on understanding mammalian systems and specifically as surrogate human systems.

"In ten years, we have seen enormous advances in understanding the immune system, in understanding hormones and their action, in understanding cancer, in understanding evolution, and even in the beginning of an understanding of the nervous system. We have seen tremendous advances in the underlying generalities of how things are organized, how genes are made, how genes are duplicated, how genes are expressed; and a side effect, one that is very significant, has been a striking unification of the kinds of problems that people in biology think about.

"If we go back ten or twenty years, hormones were thought about by physiologists, the nervous system was thought about by specialists in the various branches of the neurosciences, cancer was talked about by oncologists and physicians, evolution was discussed by population biologists, the immune system was studied by immunologists; none of that is true any longer. If you look at the seminar board on our floor at MIT, you will find seminars that cover the range of all of the things I have just talked about as well as plant biology and even the beginnings of behavioral biology, and that is a very different perspective *and has tremendous organizational consequences, and actually quite staggering implications for education and for the structure of the field.*"

We shall return in a little while to these implications. But let us first look at some other places where cross-cutting subjects are appearing, with emphasis on the study of *surface complexity arising out of deep simplicity.*

Al Scott will tell us about nonlinear systems dynamics, a very exciting branch of mathematics with applications to many parts of science. Nonlinear systems dynamics can be exemplified by first-order differential equations with several dependent variables x_i that are functions of time t. The applications are numerous. Peculiar phenomena associated with such equations come up over and over again. In the case of dissipative systems, for example, the orbits (x as a function of time) can be "attracted," at very large times, to fixed points where the x's are constant, or to limit cycles where the x's go around in periodic orbits asymptotically, or to "strange attractors" that give chaotic behavior, so that the x's at large times become infinitely sensitive to the boundary conditions on x at the initial time Chaos turns determinate systems into effectively indeterminate ones. Attempts are being made to apply these ideas to elementary particle theory, to the fascinating question of the approach to hydrodynamic turbulence, to problems of plasma turbulence, to oscillating reactions in chemistry like the Belousov-Zhabotinsky reaction, to biological clocks, and to many other areas of science.

Biological clocks, for example, seem to be nonlinear systems, each one with a free-running frequency that is usually different from what is actually needed for the clock, but with environmental signals setting the frequency as well as the phase. We are familiar with the resetting of phase in recovering from jet lag, for example. This form of clock seems to provide the kind of robustness that biological systems need Recently there have been attempts to identify the mathematical phenomena of nonlinear systems dynamics in population biology, in problems of the brain and the mind, in attempted explanations of schizophrenia, and in problems of social systems.

Many of these applications are highly speculative. Furthermore, much of the theoretical work is still at the level of "mathematical metaphor." But, I think this situation should cause us to respond with enthusiasm to the challenge of trying to turn these metaphorical connections into real scientific explanations. For that purpose, one useful advance would be to know whether these mathematical phenomena really crop up in the solution of partial differential equations, and if so, where.

There are analogous phenomena also for discrete variables, time and x. There, we connect up with fundamental areas of computer science such as Steve Wolfram has studied, including cellular automata and Turing machines. We also encounter new insights into how to construct reliable computers out of unreliable elements. I was associated more than thirty years ago with the first attempts to solve that problem, by methods that were eventually analyzed correctly by Jack Cowan. These days people are talking about more sophisticated methods, based on attractors, for making reliable computers out of unreliable elements. Attractors are almost certainly involved in this way in pattern recognition and perhaps that is true in other kinds of mental activity as well. It may be that attractors are again providing the kind of robustness that biological systems require, this time in connection with phenomena that include human thought. We might even speculate that attractors might be connected with our human habit of getting stuck in a certain way of thinking and finding it extremely difficult to jump out of the rut into another way of thinking. It would be fascinating if that turned out to be so; and understanding the situation a little better may help us to design new ways to stimulate creative thinking.

We all know that computers are not only tools for calculation, but also increasingly for symbolic manipulation, which means they can be used for doing theoretical scientific work. In many cases they can also serve as a kind of theoretical laboratory for experiments in the behavior of model systems. In addition, they are objects of study as complex systems of enormous interest in themselves. Since

World War II, in a great deal of interesting theoretical work, they have been compared with neural nets and even with human organizations. A subject embracing portions of linguistics, psychology, neurobiology, neurochemistry, computer science, and so forth, has grown up, that some people call cognitive science We all know that in most situations, theory has to advance along two tracks: the fundamental search for dynamical explanations on the one hand, and on the other, the phenomenological search for pattern in the laws of Nature. There are associated experimental domains in each case. This is true of the study of the brain, where phenomenological aspects are covered under the rubric of mind and involve the study of behavior, and sometimes, in human beings, even the study of introspection. There is always a reductionist bridge between these two kinds of explanation, the fundamental and the phenomenological. (I assume all of us are in principle reductionists.) But it often takes a very long time to construct such a bridge, such as the one between the brain and the mind, even though great strides are being made. While the construction is going on, it is necessary to pursue both approaches, which means in this case to study both the brain and the mind.

New interdisciplinary subjects are growing not only out of brain science, but also out of mental science, that is to say, psychology and psychiatry. One that I think will be of particularly great interest in the future is the scientific study of human mental processes outside awareness, what is sometimes called the unconscious mind, long dealt with by psychoanalysis, but needing to be incorporated into regular science. Pathways out of the "unconscious" are available, not only in the areas of free association, slips of the tongue, dreams and so on, but also in hypnosis and other altered states of consciousness. Hypnosis, conditioning, and, perhaps, subliminal perception may provide pathways into the "unconscious." Here, not only mental science is involved, but also physical science. With improved SQUID devices and all sorts of other tools from physical science, one may be able to discriminate by objective means among different states of consciousness, so that when we study them psychologically it will not be a circular process. As progress is made on the brain-mind bridge, the panoply of brain science or cognitive science will also be increasingly applicable. These matters will be discussed by Mardi Horowitz and Jerome Singer.

There is a striking theoretical resemblance between the process of learning and the process of biological evolution. The field of evolutionary and population biology is one to which sophisticated mathematics has been applied for a long time, with benefit to both biology and mathematics. Much has been learned and much is still not understood. We will have a brief discussion of the state of this extraordinarily important field by Mark Feldman, and we may all reflect on the benefits of future interactions among students of computers, learning, and evolution. Manfred Eigen will tell us about a laboratory system that exhibits evolutionary behavior and may be related to the chemical reactions that produced the first life on Earth; he will thus introduce us to the subject of pre-biotic chemical evolution.

Now cognitive scientists and students of various kinds of evolution are beginning to get together. A new subject is taking shape, which has roots in cognitive science, in nonlinear systems dynamics, and in many parts of the physical, biological, and even the behavioral sciences. Some people call it self-organization, others complex systems theory, others synergetics, and so forth. It tries to attack the interesting question of how complexity arises from the association of simple elements. A conference is being planned at the Center for Nonlinear Studies at Los Alamos on at least part of that new subject—the study of evolution, learning, and games, with emphasis on the theory of adaptation. The conferees will listen to reports on game theory strategies in biological evolution, the coevolution of genotype and phenotype in biological evolution, theoretical and experimental results on chemical or prebiotic evolution, the development of foraging strategies in ant colonies, strategies for the evolution of new algorithms in artificial intelligence (using crossing-over and natural selection in computer programs), models of human learning, the mathematical theory of regeneration in the visual cortex, discoveries on cellular automata and Turing machines, stability of deterrence and stability of the U.S.–U.S.S.R. arms competition, spin-glass models of neural networks, and other diverse topics. Yet the discussion is to be general, with physicists, mathematicians, population biologists, neurophysiologists, social scientists, computer scientists, and engineers all trading questions and comments. Many common threads are already evident, especially the nearly universal importance of adaptation, the need for random inputs in the search process, the importance of high dimensionality, the efficacy of recombination, and the importance of attractors and, in many cases, of numerous attractors.

Let me pick out just one topic, out of many excellent ones, to highlight as an example. John Holland, Professor of Computer Science at the University of Michigan, will describe the present state of his method for getting computers to evolve strategies for solving problems. He has a sort of community of instructions, with competition and natural selection, and variability produced by random crossing over, as in chromosomes. Lo and behold, clever new strategies emerge from his computer. So far his genetic analogy is with haploid organisms. He has not yet introduced diploid genetics—just think how much better it will be with sex.

A special subject is the evolution of human behavior, where it is evident that biological evolution has been overtaken by cultural evolution. This field has recently been enlivened by controversy between some sociobiologists, who have underestimated the cultural transformation of the biological roots of human behavior, and some cultural anthropologists, who have tried to minimize the role of biology in the explanation of human behavior. I am sure that a synthesis will emerge from this dialectic process. However, the field goes far beyond such a controversy and has contributions from paleontology, primatology, archaeology, psychology, and so forth. To consider a layman's example, some day we might be able to choose between two popular models of the evolution of organized human violence, which threatens all of us so dramatically in this era. According to one model, there has always been a tendency towards occasional intraspecific violence from early man up through the hunter/gatherer stage of culture and on to the present. As people have formed larger and larger groups, and of necessity have become organized more tightly and on a larger scale, and with improved weapons as well, the scale of violence has correspondingly grown. According to another model, somewhat different, there was a qualitative change at a certain time, perhaps at the time of the invention of agriculture, or a little later at the time of the development of hydraulic agriculture, when relatively peaceful hunter/gatherer societies were replaced by competitive societies with the concept of property. They supposedly initiated real warfare, albeit on a small scale by today's standards. Marxists tend to adopt the second model, but so do a number of thinkers who are not Marxist. It would be very interesting to be able to choose between these two ideas, or find another way of looking at the whole question. We shall hear some interesting observations from Irven DeVore and Richard Wrangham, whose studies of primate behavior bear on the possible validity of the first hypothesis.

In general, the study of prehistoric cultures now involves an intimate association of archaeology, cultural anthropology, and ecology, but physics, chemistry, botany, and many other scientific subjects are also contributing through what some people call archeometry. They mean the study of old objects, especially artifacts, by advanced technical means that can yield information not only about dates and authenticity, but also patterns of use, methods of manufacture, provenience, and, therefore, mines, trade routes, and so forth. We could in the future throw new light on the mystery of the classic Mayan collapse, for instance, which was the subject of a series of discussions here at the School of American Research some time ago that resulted in a fascinating book. Or we could understand better the successive extinctions of Pueblo cultures here in the Southwest. (I hasten to add that at the time of each extinction, some Pueblo cultures survived, and some survive to this day.) Probably, with the aid of the various disciplines working together, one can to some extent resolve these and other mysteries and, thus, understand better the conditions for the survival of human culture. On this subject, Douglas Schwartz will have some interesting insights to share with us.

In many of the areas of research we are discussing, a common element is the explosive growth of computer capability and of computer-related concepts. We have mentioned that the computer is a marvelous tool for calculations, for theoretical experimentation, and for symbolic manipulation. It is not only an aid to thinking, but a system to be studied and compared, for instance, with the brain. Undergraduate students are choosing computer science as their major subject in record numbers; they are flocking to it like lemmings. Nevertheless, some of us believe that the emerging subjects of information science and artificial intelligence are not providing a broad enough scientific and cultural foundation for research and education in the computer field. Closer ties with many fields of natural and behavioral science and with mathematics would seem to be desirable, as at the conference being planned at CNLS.

Furthermore, it is important to teach students to avoid the pitfalls of reliance on massive computer facilities. Most of us are familiar with these pitfalls. The tendency to calculate instead of to think is an obvious one: "I'll run it for you Tuesday," rather than "I'll think about it for a minute." Another tendency is to neglect essential qualitative and synthetic aspects of many systems under study in favor of mere analysis of easily quantifiable concepts. Avoiding such neglect is of great importance, because we are concerned here not only with complex physical and chemical systems and computers, but also with such subjects as language, the brain and the mind, ecosystems, and social systems and their history, for which exclusive emphasis on the analytic and quantitative aspects can be disastrous. Many of our topics link natural science, behavioral science, and the humanities, and the contribution of certain subjects in the humanities, such as history and possibly applied philosophy, may be crucial.

That is especially true in the case of policy studies. Policy studies constitute one of the most vital activities in our society, increasingly necessary for our survival. Not often discussed although widespread, policy studies concern the individual, the family, the community, the state, the nation, or even the world community. These studies consider what are the likely consequences of particular decisions; how uncertain are these consequences; how the consequences are likely to affect in some concrete way various systems of values. We have to take into account the enormous and increasing complexity of modern society. These days, much legislation, for example, accomplishes the opposite of what it sets out to promote, along with even larger and unexpected side effects. The same if often true of technical innovation, the side effects of which are notorious. A full-scale study of a local or national or world problem, properly done, would have contributions from natural science, social science, applied philosophy, (especially ethics and aesthetics), law, medicine, practical politics, and, of course, mathematics and computer science in order to handle the vast number of variables. It is very difficult to bring all these disciplines together, even in think tanks designed for that purpose. Our compartmentalization of learning is becoming more and more of a grave hazard. Here, too, it is especially important and challenging to combine mathematical sophistication in such matters with the proper consideration of value systems often difficult to quantify. Computers have exacerbated this problem, although they need not do so, and they are, of course, essential for huge studies. They need not do so, because with the aid of powerful computers, one can proceed in ways such as the following: devote great care, in any policy study, to finding really sensible surrogates or yardsticks for many of the important values involved, treating this as a major part of the work. Then, instead of assigning relative quantitative measures to the various values and simply optimizing, display in a multi-dimensional way how the the different policy options affect all those surrogates and how sensitive the effects are to changes in policy. We may find, then, for example, that minor sacrifices in one important value may allow large gains in another. It is important, of course, to estimate uncertainties as well, and even more important to use science, engineering, and general inventiveness to enlarge the sphere of policy options in order better to accommodate many important values.

Thus, we see one example of how computers can be used to render policy studies more humane. A suggestion of how mathematics teaching can accomplish something similar was the main thrust of a lecture I once gave to the students of the Ecole Polytechnique in Paris, a sort of military school of science and engineering that functions as a temple of mathematics and mathematical science for the intellectual élite of France. At that time, an invited lecturer spoke not to a few of the students, but the entire student body, which was marched in, in uniform. I started by congratulating them on being privileged to get such a splendid technical education as was offered at the Ecole, then said that, of course, many of them would end up not as scientists or engineers but as managers of great enterprises in France, and that their firm grounding in mathematics would be just as valuable there, since many sophisticated mathematical theories had been developed in economics and management. Then, to the dismay of the students and the delight of the professor of physics and the professor of social science who had invited me, I explained that what I meant was that mathematics would be useful to them defensively, so that they would not be snowed by studies in which relatively trivial matters had been quantified and carefully analyzed, while dominant values were set equal to zero for convenience. We need a balanced and humane use of mathematics in these cases, and people who have not been trained in defensive mathematics will have difficulty defending their sound qualitative judgements against the onslaught of pseudo-quantitative studies.

In my remarks so far, I have tried to sketch, with the aid of some important examples, the revolution that is taking place in science and scholarship with the emergence of new syntheses and of a rapidly increasing interdependence of subjects that have long been viewed as largely distinct. These developments pose a difficult challenge to our institutions. In my remaining time, let me discuss that challenge and one or two possible components of the response.

We have an imposing apparatus of professional societies, professional journals, university departments for research and teaching, government funding agencies, and peer review committees or sections, all directed (at least in part) toward quality control in the traditional disciplines. In the past, it has been possible to accommodate, over time and with considerable difficulty and inconvenience, but in the long run with reasonably satisfactory results, the appearance of cross-disciplinary subjects like biochemistry or nuclear engineering. I believe, however, that the current developments in science and scholarship represent a much more rapid and more widespread rearrangement of subjects than we have experienced before and that it involves much of the most important new work in science. (But certainly not all. Let me make that perfectly clear, as one of our recent national leaders used to say. I am not trying to play down the importance of individual achievement in traditional fields, which remains vital to the health of the scientific and scholarly enterprise.) The apparatus we have described needs to change more rapidly and more radically than it is accustomed to doing, and we must understand what would be useful and appropriate changes and how they might practically be carried out.

Ways will have to be found of permitting and encouraging higher education suitable for the new widely emerging syntheses. Probably in many cases longer and more varied education, perhaps even formal education, will be needed, including years of postdoctoral study and apprenticeship; and we will have to learn to adjust to the personal and economic changes involved.

The whole pattern of grants and peer review must evolve in ways that are hard to prescribe and even harder to carry out.

The journals and professional societies will have to evolve so that the establishment of standards and the conduct of refereeing can be carried out for the new transcendent subjects. All of that will be painful and difficult but exciting.

The universities will have to adjust their departmental structures and modify some of their traditional ways of selecting professors and planning curricula. Our first-class universities are in the hands of very clever people, and I am sure that gradually some suitable changes will come about, as in the other organizations, despite the existence of very considerable bureaucratic inertia. But the changes may well be slow and, for a long while, not wholly satisfactory.

Let me describe, therefore, as one important contribution to the resolution of the crisis that we face, a new institution that could serve as an example and a challenge to the older ones.

The fact that natural and social science are redefining themselves seems to create the opportunity for a new kind of institution that would combine the advantages of the open teaching and research environment of the university with the flexibility of interdisciplinary patterns in national laboratories and other dedicated research institutions.

What we propose is the creation here in Santa Fe of such an institute for research and for graduate and post-doctorate education in selected areas, based on novel principles and responsive to the trends in science and scholarship that we have just been discussing. The typical American university must provide instruction in a wide variety of fields for its undergraduates. Even an institute of technology with emphasis on science and engineering has numerous departments, especially in the humanities, that give service courses. A relatively specialized institute, such as we envision here, cannot provide the kind of general coursework that an American undergraduate is supposed to require. Such an institute should not award a bachelor's degree. Even elementary graduate instruction of the conventional kind would give rise to problems. Usually there are departments in the traditional disciplines, each offering master's as well as doctor's degrees, and each scheduling a variety of full-length lecture courses in a great many subdisciplines. Professorial staff have to be hired to attend to all those courses.

We propose a quite different structure for the new institute, and we would like to hear your comments on it. Full-scale lecture courses would not be emphasized; teaching would be accomplished mostly in seminars and short series of lectures, but, above all, by means of apprenticeship and research. Only the Ph.D. would be awarded, typically in interdisciplinary subjects forming part of the research program, although not necessarily always. Advanced graduate students would be easily accommodated in such an institution. Beginning graduate students and even occasional students without a bachelor's degree would be welcomed if they could dispense with the traditional array of long lecture courses covering the ground of each subject and dealing with material already available in books. We would hope that many of our students would have acquired as undergraduates an elementary background in natural science, mathematics, the social sciences, and some parts of the arts and humanities.

In this way we hope that the Institute can do without the usual departments. Faculty members trained in particular subfields, and with strong interdisciplinary interests of particular kinds, could be selected without worry about having all the other subfields of each particular discipline represented, because we would not try to offer a complete curriculum in that discipline.

Research groupings,which may change over time, would constitute themselves. Presumably, those research groupings would recommend to the faculty and administration highly qualified candidates for new appointments. We need your advice on how this might work. Interdisciplinary appointments, which are often so difficult to make at universities with a traditional structure, would be encouraged. At a typical university, for example, an archeometer with a Ph.D. in chemistry would have a very difficult time being appointed either to the chemistry or to the archaeology department, in one case because he is doing the wrong research and in the other case because he has the wrong degree. Most archeometers have taken refuge in other places, for example, in the basements of museums.

I had a very interesting experience a few months ago visiting a great university where there is a famous Russian research center. After a little while I found myself dragged off to see the very amiable President of the University. The Director of the Center wanted me to help him persuade the President that the University should appoint a distinguished expert in Soviet economics, who would be immensely useful to the Russian research center. He is a very good Russian scholar and a very good economist, but he was not doing what the economics department thought was its highest priority, and he was not doing what the Russian history department thought was its highest priority, and so, neither department would appoint him. I believe that ultimately common sense won out in that case, but it does not always do so.

That kind of problem is apparently very widespread. It has its foundation in a real concern that lies behind the skepticism about academics seeking interdisciplinary appointments. Faculty members are familiar with a certain kind of person who looks to the mathematicians like a good physicist and looks to the physicists like a good mathematician. Very properly, they do not want that kind of person around. In fact, our organization into professions, with professional societies, journals, traditions, and standards of criticism, has much to be said in its favor, because it helps to safeguard excellence. Presumably some new patterns of setting standards are needed, and that is something we could well discuss.

It is important to recruit for the faculty of the institute some of those rare scholars and scientists who are skilled and creative in a variety of subjects. We hope, too, that among the graduates of the Institute there would be more of this kind of person. Of course, not all the graduates would be genuine polymaths, but we would hope to turn out graduates capable not only of solving particular problems, but of thinking and analyzing and especially synthesizing in a wide variety of contexts.

Ways will have to be found of encouraging teamwork among people of the most diverse backgrounds interested in the same emerging syntheses. Here it will be important to have some scholars with synthetic minds who can grasp the similarities, especially theoretical parallels and common applicable techniques, among the many subfields under discussion and also specialists (in a few remarkable cases, the same people, but in most cases different people) who are responsibly familiar with the structure and the properties and the observational or experimental facts of each subject.

One of the challenges that we face, in tackling subjects that involve mathematics and natural science on the one hand and also social and behavioral science on the other, is that of marrying quite different intellectual cultures. The problem is exacerbated by the fact that many of the most mathematically sophisticated social scientists are those who are most attracted by the analyzable at the expense of the real. Fortunately, there are others who combine a concern with the crucial qualitative features of their subject matter with a receptivity to ideas from mathematics and natural science; and there are also natural scientists who are capable of learning about the complexities of human beings and their institutions.

There are some psychologists and pop psychologists who like to place people on a scale running from Apollonian to Dionysian, where, roughly speaking, Apollonians tend to favor logic, rationality, and analysis, while Dionysians go in more for intuition, feeling, and synthesis. In the middle are those tortured souls, the Odysseans, who strive for the union of both styles. The new institute would have to recruit a number of Odysseans to be successful!

You have read in our brochure about how we would have permanent faculty, tenure-track faculty, junior faculty, Ph.D. candidates, post-docs, visiting faculty, and nonresident fellows who would visit from time to time on a regular basis.

The research program of the Institute would include both experimental and theoretical work, which complement and reinforce each other. We would differ fundamentally, therefore, from the Institute for Advanced Study in Princeton, for example, which has no experimental work, does not award degrees (although I believe it is allowed to), and does not have very much collaboration among different kinds of scholars. Experimental and observational work of very expensive kinds, such as high energy physics, astronomy, and oceanography, should probably not be undertaken, while use is made of cooperative arrangements with nearby observatories, laboratories, museums, industrial enterprises, and so forth.

I should mention that it is very tempting to consider adding future studies and policy studies to the material covered by the Institute. There is an urgent need to apply the skills of scholars and scientists to the problems facing communities, regions, nations, and the world. However, the nature of such policy studies, along with the mix of people necessary to do justice to them, is probably sufficiently different from that of the subject we have been discussing, that it would be better (and we need your advice and comments on this) to organize an autonomous and separately funded organization nearby that would concern itself with policy studies and speculation about the future. Such a nearby think tank, if it is created, could then employ selected faculty members, visitors, and students as consultants or part-time staff members, but it would also employ a number of distinguished full-time investigators experienced in policy studies and public affairs.

We describe in the brochure how after some five to ten years of growth, the personnel of the Institute would consist of so and so many professors and so and so many secretaries, and so and so many students, but we need your advice as to whether the numbers are reasonable and how to get from here to there.

The location of the Institute in this vicinity seems to provide a uniquely attractive cosmopolitan environment in a relatively unspoiled setting. (Of course, all buildings in Santa Fe look like this one, and the weather is always the way it is today!) Recruiting a superlative faculty and gifted students will be facilitated by this choice of location. George Cowan has described the proximity of Los Alamos, the radio and optical observatories of the Southwest, the museums and the Laboratory of Anthropology in Santa Fe. There is an emerging high technology research corridor in the Rio Grande Valley. It is also remarked in our propaganda that within a thousand mile radius lie the San Francisco Bay Area, the Rocky Mountain Region, Chicago, Minneapolis, St. Louis, and all of Texas. In any case, it may be, in an age of advanced communications and satellite television, that intellectual stimulation and the exchange of ideas will not require proximity to large urban agglomerations, and that we will be pioneering in that respect as well as others.

At the same time that we will be seeking very substantial funds for the endowment and trying to work out how such an institution could best be structured and governed, we will be starting up a program of intellectual activity by establishing so-called research networks. Now research networks have a relatively long history, as exemplified fifty years ago by Delbrück and Luria, who were supported, I believe, by the Rockefeller Foundation. One does not really invent such networks; to some extent they already exist as invisible colleges, colleges without walls, but one can discover and assist them and develop them further. The MacArthur Foundation has been experimenting with such networks for the last few years, particularly in supporting research in scientific fields relevant to mental health. A subject is chosen (for example, the psychobiology of depression) and some research groups from different institutions are selected to participate in the network studying that subject. The groups and the individuals composing them represent a variety of disciplines, and the groups are chosen for the compatibility and complementarity with one another as well as for their excellence, so that they are able to function in a pattern of collaboration. The Foundation helps the groups to communicate with one another by telephone and by computer mail, by means of conferences and summer studies, and by exchanging post-docs as well as data, samples, information about methods, and so forth. It is hoped that the research network can then carry out an integrated attack on the problem it is studying.

In a somewhat analogous way, our Institute, if it can obtain operational funding, can start very soon to set up research networks for studying some of the emerging interdisciplinary syntheses we have been discussing. For each network, composed of individuals and research groups at various institutions, we will provide computer links and a budget for other kinds of communication, including meetings here in Santa Fe, probably short ones during the academic year and workshops lasting for weeks in the summer. The central headquarters here would be responsible for arranging the details.

During the early phase of operation of the Institute, there would be only a small faculty here. As academic members of the Institute begin to appear in Santa Fe (at first mostly non-resident fellows and others on leave from institutions elsewhere), there would be a few scientists and scholars representing each network locally and enhancing its cohesion.

What would be the relation of the network activity of the Institute to the existing academic and industrial organizations to which we all belong? The network activity could only be a benefit to those organizations and to their members and, if it proceeds as we hope, it would greatly facilitate the research of participants, wherever they are. At the same time, it would strengthen the nascent Santa Fe Institute. In fact, if we consider the two operations, building the networks and establishing the permanent Institute, we see that each is very valuable in itself and also that they are mutually beneficial.

As the permanent Institute gradually comes into being, there is no reason to believe that the networks will cease to operate. Assuming they are successful, they should presumably continue indefinitely and constitute one of the principal modes of operation of the Institute, adding strength to it, and also to many of the leading academic organizations in this country and to some abroad. In the long run, some of those institutions may be taxed by having one or two of their faculty members lured away, and an occasional bright student, but in exchange for that tax they would be provided with a very valuable service.

One of the most important questions that we have to address is this: Why not try to accomplish some of our objectives by adding to the activities of an existing university and saving the cost of creating a new institution? Well, I think that the national response to the challenge of the emerging syntheses *will* consist in great part of steps taken by the universities. They have already begun to respond to a considerable extent. But the form of the response, as I indicated before, is not likely to be adequate for a long time.

Let me poke a little fun at the universities and institutes of technology. The typical response of a university to the emergence of a new interdisciplinary subject is to set up a Center in an old Victorian house or a little shed left over from the First or Second World War, funded with soft money and treated to some extent like a stepchild. Wonderful results often emerge from these dilapidated structures, but some of the most talented researchers are not in permanent positions, have little influence on teaching policy, and are far removed from the centers of influence in the institution. Of course, a senior faculty member who has distinguished himself in a particular profession and made a great reputation can afford to shift to a new, interdisciplinary subject. He can sometimes get funding, although that is not very easy. However, the younger people who want to work on the new subject may have great difficulty furthering their careers, unless they wish to spend years becoming famous in some old-fashioned field.

It will be a slow and difficult process for each university to change from its old message, "Learn a traditional subject and stick to it," to the new message, "It is all right to learn how to make connections among different subjects." We would like to create here in Santa Fe at least one institution that is free from the drag exerted by past specialization and the tyranny of the departments, an institution that would encourage faculty, students and young researchers to make connections. The message, that it is all right to think about the relations among different approaches to the world, may then spread more readily to the world at large: to the universities, the technical institutes, and even to the primary and secondary schools, where innumerable opportunities to point out connections are wasted every day. Thank you.

P. W. ANDERSON
Joseph Henry Laboratories of Physics, Princeton University, Princeton, NJ 08544

Spin Glass Hamiltonians: A Bridge Between Biology, Statistical Mechanics and Computer Science

A remarkable number of fields of science have recently felt the impact of a development in statistical mechanics which began about a decade ago[1] in response to some strange observations on a variety of magnetic alloys of little or no technical importance but of long-recognized scientific interest.[2] These fields are:

1. Statistical mechanics itself, both equilibrium and non-equilibrium;
2. Computer science, both special algorithms and general theory of complexity;
3. Evolutionary biology;
4. Neuroscience, especially brain modelling;
5. Finally, there are speculations about possible applications to protein structure and function and even to the immune system.

What do these fields have in common? The answer is that in each case the behavior of a system is controlled by a random function of a very large number of variables, i.e. a function in a space of which the dimensionality is one of the large, "thermodynamic limit" variables: $D \rightarrow \infty$. The first such function of which the properties came to be understood was the model Hamiltonian

$$H = \sum_{ij}^{N} J_{ij} \, S_i S_j \tag{1}$$

(J_{ij} is a random variable, $P(J_{ij}) \propto e^{-J_{ij}^2/2(\bar{J}_{ij})^2}$, S_i a spin variable attached to si i) introduced[1] for the spin glass problem. This Hamiltonian has the property "frustration" named by G. Toulouse[3] after a remark of mine, which roughly spea ing indicates the presence of a wide variety of conflicting goals. A general definitic suitable for a limited class of applications has been proposed:[4] imagine that tl "sites" i on which the state variables reside constitute the nodes of a graph repr senting the interactions between them—simply a line for every J_{ij} in the case (1), for instance. Let us make a cut through this graph, which will have a certa area A ($\propto N^{d-1}$ in case the graph is in a metric space of dimension d). Set ea of the two halves in a minimum of its own H, normalized so that $H \propto N$. The reunite the halves and note the change in energy ΔH. If the fluctuations in Δ are of order $\sqrt{A}$, H is "frustrated"; if they can be of order A—as in (1), they w be if the J are all of the same sign—it is "unfrustrated." The dependence on $\sqrt{}$ means that when the interactions within a block of the system are relatively sati fied, those with the outside world are random in sign; hence, we cannot satisfy a interaction simultaneously.

A decade of experience with the spin glass case has demonstrated a numbe of surprising properties of such functions as H. As the Hamiltonian of a statistic mechanical system, for most dimensionalities it has a sharp phase transition in tl $N \to \infty$ $limit$[1]. At this transition it becomes non-ergodic in that different regior of phase space become irretrievably separated by energy barriers which appear 1 be of order N^p where p is a power less than unity.[5] As the temperature is lowere these regions proliferate, exhibiting an ultrametric multifurcation.[6] It is suspecte that the number of such regions at or near the minimum (ground state) of , has no entropy (not of order e^N), but may be exponential in some power of N Many unusual properties of the response functions, and some strange statistic; mechanical and hysteretic behaviors, have been explored at length. Recent wor has generalized the Hamiltonian (1) and also shown that even first-order pha: transitions may occur for some models.

In computer science, there are a number of classic optimization problems whic have been studied both as objects for heuristic algorithms and as examples for con plexity theory. These include the spin glass itself (sometimes under other names the graph partition problem (which can be transformed into a spin glass), graph co oring (close to a Potts model spin glass), the Chinese postman (in some cases equi alent to a spin glass) and the famous travelling salesman (Design a tour throug N cities given distances d_{ij}, of the minimum length $L = \sum d_{ij}$). As I indicate several of these are spin glasses—there even exists a very inefficient transformatic due to Hopfield[7] $TS \iff$ spinglass—and all are well-known NP-complete—i. hard—problems, of which it is speculated that no algorithm will solve the gener; case faster than $O(e^N)$.

On the level of heuristic algorithms, Kirkpatrick[8] has suggested that the pro cedure of annealing using a Mitropolis–Teller Monte Carlo statistical mechanic algorithm may be more efficient for some of these problems than the convention; heuristics. In any case, the knowledge that a "freezing" phase transition exists an that, for values of the function below freezing, one may be stuck forever in an unfavorable region of phase space, is of great important to the understanding of the structure of such problems. To my knowledge, the computer science community only knew of freezing as a bit of folklore, and has not yet absorbed its fundamental importance to the whole area—which includes great swatches of problem-solving and AI. Incidentally, a workshop at BTL came up with the limited but interesting conclusions that (a) simulated annealing works; (b) sometimes—not always—it beats previously known heuristics.[9]

Equally important should be the knowledge that there is a general theory of *average* properties of such problems, not limited to the mathematicians' type of worry over worst cases, but able to make statements which are overwhelmingly probable. For instance, we also know analytically the actual minimum energy to order N for "almost all" cases of several kinds of spin glass. A student (Fu) and I have an excellent analytic estimate for the partition problem on a random graph, etc. We also can hope to achieve a real connection between algorithmic solution and non-equilibrium statistical mechanics: after all, the dynamic orbits of a system are, in some sense, the collection of all paths toward minimum energy, and, hence, of all algorithms of a certain type (D. Stein is working on this).

In evolutionary biology, we can consider the fitness—the "adaptive landscape" as a function of genome to be just the kind of random function we have been talking about. The genome is a one-dimensional set of sites i with 4 valued spins (bases) attached to the sites, and the interaction between the different sites in a gene is surely a very complicated random affair. In the work of Stein, Rokhsar and myself,[10] we have applied this analogy to the prebiotic problem, showing that ·it helps in giving stability and diversity to the random outcomes of a model for the initial start-up process. Here we see the randomness as due to the tertiary folding of the RNA molecule itself.

G. Weisbuch has used a spin glass-like model for the evolutionary landscape[11] to suggest a description of speciation and of the sudden changes in species known as "punctuated equilibrium." Most population biology focuses on the near neighborhood of a particular species and does not discuss the implications of the existence of a wide variety of metastable fixed points not far from a given point in the "landscape."

In neuroscience, Hopfield[12] has used the spin-glass type of function, along with some assumed hardware and algorithms, to produce a simple model of associative memory and possibly other brain functions. His algorithm is a simple spin-glass anneal to the nearest local minimum or pseudo-"ground state." His hardware modifies the J_{ij}'s appropriately according to past history of the S_i's, in such a way that past configurations $\vec{S} = \{S_i\}$ are made into local minima. Thus, the configuration $\{S_i\}$ can be "remembered" and recalled by an imperfect specification of some of its information.

Finally, for our speculations for the future. One of these concerns is biologically active, large molecules such as proteins. Hans Frauenfelder and his collaborators have shown that certain proteins, such as myoglobin and hemoglobin, may exist in a large number of metastable conformational substates about a certain tertiary structure.[13] At low temperatures, such a protein will be effectively frozen into one of its many possible conformations which in turn affects its kinetics of recombination with CO following flash photolysis. X-ray and Mössbauer studies offer further evidence that gradual freezing of the protein into one of its conformational ground states does occur. Stein has proposed a spin glass Hamiltonian to describe the distribution of conformational energies of these proteins about a fixed tertiary structure as a first step toward making the analogy between proteins and spin glasses (or possibly glasses) explicit. In any case, this field presents another motivation for the detailed study of complicated random functions and optimization problems connected with them. Yet another such area is the problem of the immune system and its ability to respond effectively to such a wide variety of essentially random signals with a mechanism which itself seems almost random in structure.

REFERENCES

1. S. F. Edwards and P. W. Anderson, *J. Phys.*, F. 5, 965 (1975).
2. Original observation: Kittel Group, Berkeley, e.g. Owen, W. Browne, W. D. Knight, C. Kittle, *PR* **102**, 1501-7 (1956); first theoretical attempts, W. Marshall, *PR* **118**, 1519-23 (1960); M. W. Klein and R. Brout, *PR* **132**, 2412-26 (1964); Paul Beck, e.g. "Micto Magnetism," *J. Less-Common Metals* **28**, 193-9 (1972) and J. S. Kouvel, e.g. *J. App. Phys.* **31**, 1425-1475 (1960); E. C. Hirshoff, O. O. Symko, and J. C. Wheatley, "Sharpness of Transition," *JLT Phys* **5**, 155 (1971); B. T. Matthias et al. on superconducting alloys, e.g. B. T. Matthias, H. Suhl, E. Corenzwit, *PRL* **1**, 92, 449 (1958); V. Canella, J. A. Mydosh and J. I. Budnick, *J.A.P.* **42**, 1688-90 (1971).
3. G. Toulouse, *Comm. in Physics* **2**, 115 (1977).
4. P. W. Anderson, *J. Less-Common Metals* **62**, 291 (1978).
5. N. D. Mackenzie and A. P. Young, *PRL* **49**, 301 (1982); H. Sompolinksy, *PRL* **47**, 935 (1981).
6. M. Mézard, G. Parisi, N. Sourlas, G. Toulouse, M. Virasoro, *PRL* **52**, 1156 (1984).
7. J. J. Hopfield and D. Tank, Preprint.
8. S. Kirkpatrick, C. D. Gelatt and M. P. Vecchi, *Science* **220**, 671-80 (1983).
9. S. Johnson, private communication.
10. P. W. Anderson, *PNAS* **80**, 3386-90 (1983). D. L. Stein and P. W. Anderson, *PNAS* **81**, 1751-3 (1984). D. Rokhsar, P. W. Anderson and D. L. Stein, Phil. Trans. Roy. Soc., to be published.
11. G. Weisbuch, *C. R. Acad. Sci. III* **298** (14), 375-378 (1984).
12. J. J. Hopfield, *PNAS* **79**, 2554-2558; also Ref. 7.
13. H. Frauenfelder, *Helv. Phys. Acta* (1984) in press; in *Structure, etc.*

MANFRED EIGEN
Max-Planck-Institut für biophysikalische Chemie,3400 Göttingen, Federal Republic of Germany

Macromolecular Evolution: Dynamical Ordering in Sequence Space

ABSTRACT

Evolution of self-replicating macromolecules through natural selection is a dynamically ordered process. Two concepts are introduced to describe the physical regularity of macromolecular evolution: sequence space and quasi-species. Natural selection means localization of a mutant distribution in sequence space. This localized distribution, called the quasi-species, is centered around a master sequence (or a degenerate set), that the biologist would call the wild-type. The self-ordering of such a system is an essential consequence of its formation through self-reproduction of its macromolecular consti tuents, a process that in the dynamical equations expresses itself by positive diagonal coefficients called selective values. The theory describes how population numbers of wild type and mutants are related to the distribution of selective values, that is to say, how value topography maps into population topography. For selectively (nearly) neutral mutants appearing in the quasi- species distribution, population numbers are greatly enhanced as compared to those of disadvantageous mutants, even more so in continuous domains of such selectively valuable mutants. As a consequence, mutants far distant from the wild type may occur because they are produced with the help of highly populated, less distant precursors. Since values are cohesively distributed, like mountains on earth, and since their positions are multiply connected in the high-dimensional sequence space, the overpopulation of (nearly) neural mutants provides guidance for the evolutionary process. Localization in sequence space, subject to a threshold in the fidelity of reproduction, is steadily challenged until an optimal state is reached. The model has been designed according to experimentally determined properties of self-replicating molecules. The conclusions reached from the theoretical models can be used to construct machines that provide optimal conditions for the evolution of functional macromolecules.

Keywords: Molecular Quasi-species, Value Topology, Mutant Population, Optimization, Evolution Experiments

1. LIFE, A DYNAMICALLY ORDERED STATE

A living system is the prototype of a highly complex, dynamically ordered state. In view of its complexity we are led to ask about the way in which such an ordered state could achieve optimal functional efficiency. The main point of my contribution will be to indicate that optimization in biology is a physical regularity associated with natural selection. It is not something that just occurred accidentally. There are principles related to precise physical conditions under which optimization of complex, dynamically ordered states is possible.

The complexity we want to consider appears already at the lowest functional level in molecular biology. Let us focus on a small protein molecule made up of a hundred amino acid residues. Twenty classes of natural amino acids account for $20^{100} \approx 10^{130}$ alternative sequences of this length that involve quite a spectrum of different functional types and efficiencies. In fact, the huge majority of these sequences does not represent any useful function. Nevertheless, the set as a whole includes all possible functions that are typical for proteins in living organisms and many more that have not yet been materialized in nature. Numbers as large as 10^{100} or 10^{1000} simply escape our capability of imagination. The whole universe comprises a mass equivalent to "only" 10^{80} protons or to about 10^{76} protein molecules of the mentioned size.

Considering any enzyme molecule, we encounter usually optimal catalytic performance. Each single physical step fits into the complex overall mechanisms so as to allow the reaction to proceed with the highest possible rate. This performance represents an optimal compromise between specificity and speed, or in other words, between selective binding of the substrate and its turnover including association and dissociation of both substrate and product. For instance, the turnover numbers of some hydrolytic enzymes reach orders of magnitude as high as 10^5 to 10^6 $[sec^{-1}]$. In such reactions, protons which must dissociate from acidic groups having pK values around 7 are shuffled around. Recombination of protons with such groups is diffusion controlled, yielding rate constants of 10^{10} to 10^{11} $[M^{-1}\,sec^{-1}]$. For pK-values of 7, the maximum rate constants for dissociation then are of the order of

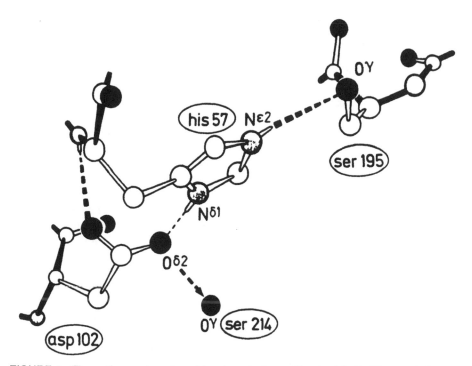

FIGURE 1 The active centre of α-chymotrypsin (according to ref. 2). The catalytic charge relay system includes hydrogen bonds between O^γ (ser 195) and $N^{\epsilon 2}$ (his 57) as well as between $N^{\delta 1}$ (his 57) and $O^{\delta 2}$ (asp 102), the numbers referring to the positions in the polypeptide chain. The centre is further fixed by hydrogen bonds between $O^{\delta 2}$ (asp 102) and O^γ of a serine at position 214 and between $O^{\delta 1}$ (asp 102) and the amido-N of his 57. The three points specified by O^γ (ser 195), $O^{\delta 2}$ (asp 102), and the center of the imidazole ring (his 57) form a plane which also contains the anticipated position of the peptide or ester substrate to which a proton is to be transferred in the catalytic process. The same charge relay system is found in subtilisin, the primary sequence of which shows no kinship relation to that of chymotrypsin. The subtilisin site involves a hydrogen bond between O^γ of ser 221 and $N^{\epsilon 2}$ of his 64, another one between $N^{\delta 1}$ of his 64 to $O^{\delta 2}$ of asp 32 and a third one between $O^{\delta 2}$ of asp 32 and O^γ ser 33. These atoms again lie in a plane with the imidazole ring (deviations < 0.2 Å). The similarity to the H-bond network at the active centre of chymotrypsin is evident, but the sequence positions of the involved amino acid residues are entirely different, and the lack of homology between the two sequences indicates convergent evolution to an optimal active centre.

magnitude of 10^3 to 10^4 $[sec^{-1}]$. H-bond connections among cooperating side chains may speed up these rates by one to two orders of magnitude. Hence, overall turnover numbers of 10^5 to 10^6 sec^{-1} indeed represent an upper limit of physically possible efficiency.

Before approaching the problem of optimization we have to ask whether there is really only one optimal state for any functional protein. The answer is clearly no. Sequence analysis of proteins at various phylogenetic levels (e.g., cytochrome c[1]) reveals differences that in many cases exceed 70%. Thus a given enzyme optimal in one organism can differ in more than 70% of its amino acid residues from another enzyme catalyzing the same reaction most efficiently in a different organism.

Another striking example is the independent evolution of the same active site in entirely unrelated sequences, as is known for chymotrypsin[2] (figure 1), a pancreatic enzyme typical of higher organisms, and subtilisin, a proteolytic enzyme produced by microorganisms, such as Bacillus Subtilis.[3] There is no sequence homology among these two enzymes and yet the two different chains are folded so as to yield practically identical charge relay systems at their active sites. Hence the optimal solution to a catalytic problem has been achieved twice in the same manner, yet via entirely independent evolutionary routes.

Moreover, site-directed mutagenesis as effected by modern genetic technology has opened a way to study systematically the functional consequences of substitutions of amino acids in a given enzyme. It turns out that many of the produced mutant phenotypes are as efficient as the wild-type enzymes, the loss in "functional value" being within a few percent.

This kind of continuity in value distribution is something we are familiar with in our surroundings. If we consider the heights of geometrical points on the surface of the earth, we see continuity rather than random distribution. Mainly we find connected planes and mountainous regions, and only in a very few places do heights change (almost) discontinuously. By analogy, if one would change the histidine at position 57 in chymotrypsin (figure 1), one would damage sensibly the active site and probably lose all functional efficiency, whereas changes at many positions outside of the active site will be of much lesser consequence. The value landscape of proteins is related to the folding of the polypeptide chain and—except for a few strategic positions— sequence similarities will also map into functional similarities. Thus, in the landscape of values, we must be aware of the presence of many peaks that are interconnected by ridges. Yet there remains a problem for evolutionary optimization. Let us assume that by optimal folding of different polypeptide chains one could produce a large number of different enzyme molecules that are virtually equally efficient in catalyzing a given reaction. However large this number may be on absolute grounds, it will be negligibly small as compared to numbers of the order of magnitude 10^{100} to 10^{1000}. In order to reach any of these optimal sequences by starting from random precursors one needs guidance; otherwise one gets hopelessly lost in the huge space of mutants. Darwin's principle of natural selection explains the existence and prevalence of optimal enzymes, but it does not yet show how optimization actually was achieved in nature.

2. THE CONCEPT: SEQUENCE SPACE

In view of the hyperastronomical orders of magnitude of possible sequences, we need an appropriate space for their representation. Our three-dimensional geometrical space is neither sufficient to accommodate within reasonable limits such big numbers nor does it offer any suitable way of representing correctly the kinship relations among the various sequences. What we need is a space that allows us to construct continuous evolutionary routes in which kinship distances (i.e., Hamming distances between related genotypes) are correctly reflected. How this can be achieved is shown in figures 2 and 3. We thus need a point space where the number of dimensions corresponds to the number of positions in the sequences which here for simplicity are assumed to be of uniform length. For binary sequences each coordinate consists of two points assigned to the two alternative binary digits. For the sequence space of nucleic acids, each coordinate is to be assigned four equivalent points representing the four possible occupations: G, A, C, T (or U). This concept of representing genotypes by a point space was first introduced by I. Rechenberg[4] and later applied also by R. Feistel and W. Ebeling.[5]

What do we gain by such an abstract concept? Apart from the fact that only in this high-dimensional space are mutant distances correctly represented, we realize that despite the huge number of states that can be accommodated in such a space, distances remain relatively small. Moreover, as figure 3 shows, the connectivity

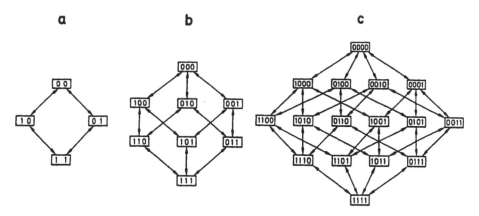

a   **b**   **c**

FIGURE 2   The correct representation of kinship distances among sequences of length ν can only be achieved in a ν-dimensionalpoint space. Three examples of binary sequences are shown: (1) $\nu = 2$, (b) $\nu = 3$, (c) $\nu = 4$. If four digit classes (e.g., four nucleotides) are involved, each axis has four equivalent positions. Case b ($\nu = 3$) then would be represented by a 4x4x4 cube comprising 64 state points.

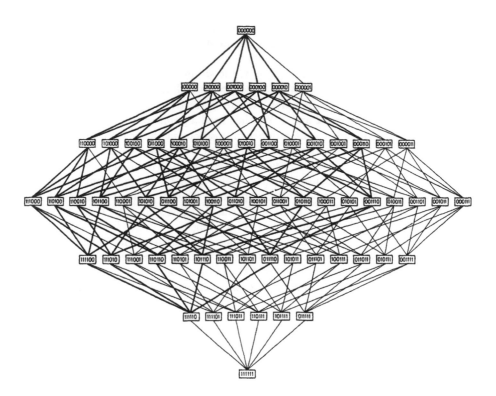

FIGURE 3 The enormous increase of connectivity among states with increasing dimension is shown for a binary sequence with $\nu = 6$ positions. The fat lines refer to a sequence with only 5 positions. While the number of states increased with 2^ν, the number of shortest possible routes between the two extreme states increases with $(\nu!)$. The diagram may be viewed also as a mutant region within a high-dimensional sequence space (ν large). In this space there may exist highly interconnected regions of this kind that show high selective values and therefore become preferentially populated.

among the various states greatly increases with increasing dimension. Hence the chances to get stuck on a local peak greatly diminish, especially if jumps (i.e. multiple mutations) are possible.

The enormous reduction of distances at the expense of a (moderate) increase in dimensions pays off only if biassed random walk processes (i.e. processes that are guided by gradients) are to be dealt with. This may be seen from the example represented in figure 4, where an unbiassed diffusion-controlled encounter

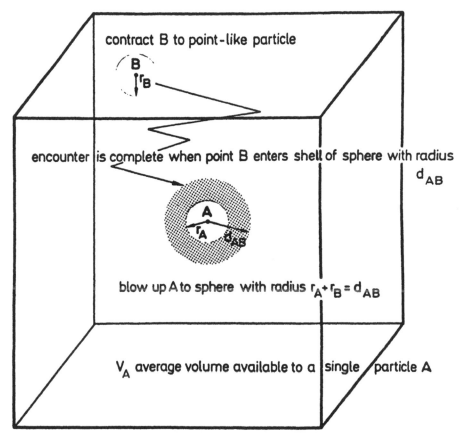

FIGURE 4 Encounter between particles A and B through random walk. Particle A is fixed in the centre of its territory, i.e., the volume V_A available to each single particle A (given by the reciprocal particle concentration of A). Particle B then describes a motion relative to A that is characterized by the sum of both diffusion coefficients: $D_A + D_B = D_{AB}$. The distance between A and B at an encounter is $d_{AB} = r_A + r_B$ where r_A and r_B are the radii of the sphere-like particles A and B. One may describe the encounter by the flux of point-like particles B into a sink which is represented by a sphere with the radius D_{AB}. The stationary solution of the diffusion equation in polar coordinates yields $n_B/n_B = 4\pi D_{AB} d_{AB}/V_A$. D_{AB} may be expressed as $d_A^2 B/6\tau_{AB}$ where τ_{AB} then is the average time required for scanning the volume $v_{AB} = 4\pi d_{AB}^3/3$ by diffusional motion. The average time required for a single encounter between the two particles A and B then is $2\tau_{AB} V_A/v_{AB}$, i.e. $2\tau_{AB}$, the time involved in scanning the volume element v_{AB} by 3-dimensional diffusional motion, times the number of such volume elements that fit the total territory of A, the volume V_A.

between two particles A and B is considered. The time required for each particle B to encounter a target particle A is given by twice the time required for diffusional motion over the encounter distance d_{AB} (i.e. the radius of the spherical sink within which A and B have to meet for an encounter to be complete) times the ratio of the two volumes, i.e. the total volume V_A to be scanned by diffusive motion of particle B and the target volume v_{AB}. This ratio, so to speak, is the number of volume elements which on average have to be scanned through by particle B before the target is reached. If the target were a lattice point and if motion were effected through hopping among lattice points, the time is simply twice the time required for hopping from one point to the next times the number of lattice points of the territory of each particle A. This trivial result also applies to higher dimensional spaces. In an unbiassed random walk process, the time to reach a particular point is proportional to the number of points that have to be scanned. Hence reduction of distances through increase of dimension is of no help in this case.

Selection, however, is a biassed process, as will be shown in the following sections. For such a biassed process, decrease of distance and increase of connectivity are of great assistance in reaching a particular target. Even in the unbiassed random walk, it doesn't take much time for the system to travel over relatively large distances and thereby get into an entirely new environment. Our task is now to find out how sequence populations are distributed among sequence space, how they localize and how they approach particular targets.

3. THE CONCEPT: QUASI-SPECIES

At the time when Darwin formulated his principle of natural selection as "survival of the fittest," there was no way to define "fittest" other than by the fact of survival. Therefore one argued that Darwin's principle might be a mere tautology, i.e., "survival of the survivor." Population genetics in the first half of this century corrected this misconception by showing that competitive growth can be formally described by means of differential equations, the solutions of which can simulate "natural selection." In these equations a combination of dynamic parameters appears which is decisive for the outcome of competition and which therefore was called "selective value." This quantity is related to inheritable properties which express themselves phenomenologically through fecundity and mortality of the individual characters.

Nevertheless, Darwin's principle in its simple form poses a problem. It postulates a relation between population parameters (survival means non-zero population numbers) and a value parameter (as expressed by the term "fittest"). It is this correlation between population and value topology that we have to analyze in more detail. How is value topology mapped into population topology? Does this correlation provide a continuous route of evolution to the highest peaks in value space?

The biologist usually identifies with the term "fittest" the wild type of a given population. If—even under ideal selection conditions—the wild type were the only survivor, we would run into a serious problem with regard to optimization. In a fairly early stage, the evolutionary process would have stopped on a quite minor foothill in value space. The existence of any monotonically rising route from such an initial foothill (corresponding to an initially poorly-adapted wild type) to any of the high mountain regions in value space that correspond to an optimally adapted phenotype (as encountered in any present living organisms) would not be very likely.

Apparently we must look more closely at the population numbers in the mutant distribution which might provide guidance for further adaptation. Mutants are at first produced with frequencies that correspond to their kinship relation to the wild type. These mutants, however, do reproduce themselves, and their population numbers finally depend not only on their kinship relation to the wild type but also on their own selective values relative to that of the wild type. These mutants, if appearing in large numbers, will again produce mutants that are rated according to their selective values. Any cohesive structure of this distribution then will provide guidance for the evolutionary process.

Before discussing the consequences of this guidance by cohesive value landscapes, let me review in a comprehensive form the mathematical formalism of a concept we have called: quasi-species.[6]

Let n_i be the population number of species i and w_{ii} a positive diagonal coefficient which implies that species i during its lifetime produces an excess of entirely correct copies of itself (i.e., it reproduces correctly faster than it dies). In addition, species i is formed, through erroneous copying, by closely related mutants k. The magnitude of the off-diagonal coefficients w_{ik} depends on the reproduction rates of the mutants k and their kinship distances to species i. Close relatives therefore will have large, corresponding, off-diagonal coefficients while distant relatives will have comparatively small ones. The rate equations then read:[7]

$$\dot{n}_i(t) = w_{ii}n_i(t) + \sum_{k \neq i} w_{ik}n_k(t) + \emptyset_i \tag{1}$$

$\emptyset_i$ being a flow term controlling the constraints. We are interested in the relative population numbers of species:

$$x_i(t) = \frac{n_i(t)}{\sum_k n_k(t)}. \tag{2}$$

Accordingly the derivative of the sum $(\sum_k n_k(t))$ enters eq. (1) defining an average productivity

$$\overline{E}(t) = \frac{\sum_i \sum_k w_{ik}n_k}{\sum_k n_k} \tag{3}$$

If the flow term $\emptyset_i$ is linearly related to x_i, this term simply drops out.

Hence in relative population coordinates, eq. (1) reads:

$$\dot{x}_i(t) = \left\{ w_{ii} - \overline{E}(t) \right\} x_i(t) + \sum_{k \neq i} w_{ik} x_k(t) \tag{4}$$

This equation contains a threshold term that is given by the average production $\overline{E}(t)$. Furthermore, according to the definition of x_i, we have $\sum_k x_k(t) = 1$ and $\sum_k \dot{x}_k(t) = 0$. Eq. (4) is inherently nonlinear since $E(t)$ depends on all x_i. The threshold nature already shows that selection as a relative redistribution of the population results as a consequence of self-replication (i.e. of the existence of a positive term $w_{ii} x_i$.

B. L. Jones et al.[8,9] have shown that using the time dependent transformation

$$z_i(t) = x_i(t) exp \left(\int_o^t \overline{E}(\tau) d\tau \right), \tag{5}$$

eq. (4) can be written in the linear form

$$\dot{z}_i(t) = \sum_k w_{ik} z_k(t) \tag{6}$$

By means of eq. (5) and the solutions of eq. (6), one can construct a set of variables y_i (with $\sum_k y_k(t) = 1$), which are some kind of normal modes of the x-variables and which evolve according to

$$\dot{y}_i(t) = \left\{ \lambda_i - \overline{\lambda}(t) \right\} y_i(t) \tag{7}$$

with λ_i being the eigenvalues of the matrix $W = (w_{ij}$ and $\overline{\lambda}(t)$ being their average

$$\overline{\lambda}(t) = \frac{\sum_k \lambda_k y_k(t)}{\sum_k y_k(t)} = \overline{E}(t). \tag{8}$$

All modes belonging to an λ_i larger than the average $\overline{\lambda}(t)$ will grow up while those with $\lambda_i < \overline{\lambda}(t)$ will die out, thereby shifting the average (8) to larger values until it finally matches the maximum eigenvalue λ_m in the system:

$$lim_{t \to \infty} \overline{E}(t) \longrightarrow \lambda_m \tag{9}$$

In most cases the largest eigenvalue will be the maximum diagonal coefficient up to a second-order perturbation term $\sum_k (w_{mk} w_{km} / w_{mm} - w_{kk})$ and corresponding higher order terms.

The extremum principle (9) supposes the existence of a largest eigenvalue λ_m. Let the mean copying accuracy of a single digit be $\overline{q}$ and let the sequences be comprised of ν nucleotides. Then each generation will produce only the fraction $\overline{q}^\nu$ of the correct copies ($\overline{q}$ originally is a geometric mean which matches the arithmetic mean if all individual q-values are sufficiently close to one). The wild type m to which the maximum eigenvalue (λ_m) refers has to be more efficient in reproduction in order to make up for the loss caused by error production. Otherwise the error copies would accumulate and cause the wild- type information to disappear. Hence a threshold relation for the copying fidelity exists which can be written as

$$\bar{\sigma}_m \bar{q}^{\nu} > 1. \tag{10}$$

The precise form of $\bar{\sigma}_m$ follows from relation (9). In the simplest case (homogeneous error rate, negligible death rate) $\bar{\sigma}_m$ is the ratio of the wild-type reproduction rate to the average reproduction rate of the rest. It follows immediately from eq. (10) that there is also a threshold for a maximum sequence length: ν_{max} for which the important relation

$$\nu_{max} = \ln \bar{\sigma}_m / (1 - \bar{q}_m) \tag{11}$$

holds. The information content of a stable wild-type is restricted, the upper bound being inversely proportional to the average single-digit error rate $(1 - \bar{q}_m)$.

If, for instance, a polynucleotide chain reproduces with 1% error rate, then the sequence must not be much larger than about one hundred nucleotides (depending on $\ln \bar{\sigma}_m$) in order to preserve its information indefinitely.

The above relations are based on the validity of the approximations of second-order perturbation theory. How small the higher order terms actually are, may be demonstrated with a simple example. Consider a homogenous constant error rate $1 - q$ and homogenous replication rates $w_{kk} < w_{mm}$. The probability to produce a mutant k having d substitutions ($d = d_{mk}$ being the Hamming distance between wild type m and mutant k) amounts to

$$Q_d = \binom{\nu}{d} q^{\nu-d}(1-q)^d. \tag{12}$$

The error class belonging to the Hamming distance d comprises

$$N_d = \binom{\nu}{d}(\kappa - 1)^d \tag{13}$$

different sequences ($\kappa = 4$ is the number of digit classes, i.e., the four nucleotides of A, U, G and C).

Hence the probability to produce a given mutant copy with Hamming distance d is:

$$P_d = \frac{Q_d}{N_d} = q^{\nu}\left\{\frac{q^{-1}-1}{\kappa - 1}\right\}^d = q^{\nu}p^d \tag{14}$$

The second-order perturbation expression of λ_m then can be written as

$$\lambda_m = w_{mm} + \sum_k \frac{w_{mk} w_{km}}{w_{mm} - w_{kk}} = w_{mm} \left(1 + \sum_{d=1}^{\nu} \binom{\nu}{d} \frac{(q^{-1} - 1)^{2d}}{(\kappa - 1)^d (\sigma - 1)} \right) \qquad (15)$$

($\sigma = w_{mm}/w_{kk}$, where w_{kk} has been assumed the same for all mutants). For all cases in which the total sum is small compared to 1, it can be replaced by its first term (i.e., $d = 1$). This term then amounts to $\nu(q^{-1} - 1)^2/(\kappa - 1)(\sigma - 1)$, which for $(1 - q) \approx 1/\nu$ (error rate $(1 - q) \approx (q^{-1} - 1)$ adapted to threshold $\nu \approx \nu_{max}$) leaves with $\kappa = 4$: $1 - q/3(\sigma - 1)$. Hence the approximation is valid if the error rate remains small compared to $\sigma - 1$, or, with an error rate of 10^{-3}, adapted to the information content of a typical gene, an average advantage of the wild type over its mutant spectrum of only one percent in replication rate would be sufficient for the approximation to be valid. The second-order perturbation theory result therefore should apply to most practical cases, especially if mutant distributions are localized in sequence space.

However, there is a principal difficulty involved in the above treatment if it is applied to an evolutionary process. Assume we have a continuous distribution of selective values $f(w)$ for the mutant spectrum. Then the fitness values of interest are just those which are close to or even identical with that of the wild type, since evolution will proceed through those mutants, including neutral ones, and it is very likely that there are almost continuous routes via those mutants up to the selectively advantageous copies. These, upon appearance, will violate the threshold relation and destabilize the former wild type. A neutral mutant, although it may be quite distant from the wild type, will violate the convergence of the perturbation procedure due to the singularities resulting from denominator terms $w_{mm} - w_{kk} = 0$. J. S. McCaskill[10] therefore has extended the deterministic analysis of the quasi-species model by renormalization of the higher-order perturbation solutions. He showed that also for continuous distributions of replication rates, a localization threshold exists that is independent of population variables and confirms relation (11) for the appropriate conditions of application of second-order perturbation theory. Moreover, it predicts a localization of a stable distribution in the space of mutant sequences even in the presence of mutants arbitrarily close in exact replication rate to the maximum, i.e., to the wild-type value. The threshold relation, of course, depends on the nature of the distribution $f(w)$. It has, however, the general form of eq. (11), where $\ln \bar{\sigma}_m$ now is replaced by a term which is typical for the assumed distribution and number of sequences sampled, but independent of population variables. In any case for physically reasonable distributions, the term replacing $\ln \bar{\sigma}_m$ (according to J. S. McCaskill's estimates) remains below five or six.

4. SOME EXPERIMENTAL RESULTS

It may be worthwhile to mention now some experiments which confirm the relations given above. Those experiments have been carried out under appropriate conditions for which the quasi-species model is valid.

Ch. Weissmann and coworkers[11] by site-directed mutagenesis prepared defined mutant RNA sequences of the genome of the bacteriophage Q_β, an RNA virus that uses E. coli as its host. Comparative measurements of the replication rates of the mutants and the wild type and determination of the time lag for revertant formation in vitro and in vivo allowed them to determine the error rate.[12] They found a value of 3×10^{-4} for $(1 - \bar{q})$ and one may estimate from their data a $\bar{\sigma}_m$-value of 4. The determined length of 4200 nucleotides for the genome of the phage Q_β[13] then lies within the error limits of the threshold length $\nu_{max} = 1n\,4/3 \times 10^{-4} \approx 4500$ nucleotides. The theory states that, if the actual length is close to the threshold length, wild type becomes only a minor fraction of the total population. By cloning single mutants and determining their fingerprints, Weissmann and coworkers could show that these implications of theory are fulfilled. They found wild type to be present to an extent of less than about 5% of the total population.[11]

Competition experiments between variants of RNA molecules that can be replicated by the enzyme Q_β-replicase have been carried out by Ch. Biebricher, et al.[14] Those experiments were based on careful kinetic studies establishing the range of exponential growth to which the ansatz of the quasi-species model applies. Kinetics were studied experimentally, by computer simulation and by analytical theory. The results are published in detail elsewhere.[15,16] These studies on real self-replicating macromolecules confirm the essential kinetic properties on which the quasi-species model was based. The data show quantitatively how selection among unrelated species occurs, and that the resulting survivor builds up a mutant spectrum, as is suggested by the model. An evaluation of data which were obtain by S. Spiegelman and coworkers[17,18] shows that mutants with selective values close to that of the wild type do exist and cause strong biasses on the population numbers in the mutant distribution.

5. THE OPTIMIZATION PROBLEM

Before going on, let us briefly review the situation. We have started from the concept of a ν-dimensional sequence space, a space of points each of which represents one of the 4^ν possible polynucleotide sequences of length ν in such a way that kinship distances among all sequences are correctly represented. Each of the sequences is characterized by a selective value, a combination of kinetic constants that describes how efficiently a particular sequence reproduces and thereby is conserved in the evolutionary competition. The selective values, of course, depend on environmental conditions. Given the distribution of selective values, we have then asked how sequences are populated accordingly and, in particular, what the conditions for localization of a distribution in sequence space are.

For that purpose we have introduced the concept of a concentration or population space made up of coordinates that relate to the relative population variables x_i of various sequences. In fact, in the deterministic model we consider only the subspace of those sequences which are present with non-zero population numbers and for which the kinetic equations are to be solved. It was shown that under certain conditions such a distribution starting from arbitrary initial population numbers approaches a steady state corresponding to a localized distribution in sequence space centered around the most viable sequence or a degenerate set of such sequences. Localization depends on the maintenance of a threshold condition which can be violated by the appearance of a more viable new sequence. The (meta-)stable localized distribution was called a quasi-species.

Note that with this procedure we have only answered the question how a particular value distribution under constant environmental conditions maps into a population distribution. Localization of the population in sequence space may be called "natural selection," the target of which is not one singular sequence any more but rather the particular distribution we have called a quasi-species. This difference from the usual interpretation of natural selection will turn out to be instrumental in solving the optimization problem. If natural selection were to mean that essentially only the wild type is populated while mutants occasionally appear on a more or less random basis, the evolutionary process indeed would get stuck in the local environment of a value hill for which the distance to the next higher hill may be too large to be bridged by random mutation. If, on the other hand, natural selection includes, besides the wild type, mutants that are fairly far apart from the wild type and, if these are potential precursors of better adapted wild types, then guidance of the evolutionary process in the direction of favourable mutants becomes possible. It is immediately seen that such guidance would require first that mutant population critically depends on the various selective values and that distribution of selective values in sequence space is not completely random but rather somewhat cooperative or cohesive.

The next step therefore is to analyze in some more detail the population structure of a mutant distribution. Eq. (14) describes the probability P_d according to which a particular mutant with Hamming distance "d" is produced by the wild type. If those mutants did not reproduce themselves, the mutant distribution would simply have this form corresponding to a Poissonian (or more precisely: a binomial) distribution of the mutant classes according to eq. (12). As was shown with the example of a gene consisting of about 10^3 nucleotides, the probability for producing, e.g., a particular three-error mutant would have dropped already to values below 10^{-10}. (P_d drops by a factor of about 3×10^{-4} for any one unit of increase in the Hamming distance.) This is mainly a consequence of the huge increase of different mutant copies with Hamming distance d. The total number of mutants produced for each class d essentially drops only with ($d!$). Hence one usually will find still some high-error mutants (e.g., up to $d = 15$) in any distribution (as typical for laboratory conditions) despite the fact that the probability for any individual copy has dropped to exceedingly low values (e.g., $P_d < 10^{-52}$ for $d = 15$ in case of the above example). The important point is that the few high-error copies still populated will be those which have selective values quite close to that of the wild type.

If the mutants are supposed to reproduce themselves according to their individual selective values w_{ii}, the population distribution gets drastically modified. According to the second-order perturbation approximation presented above, the fraction of relative population numbers of mutant i in error class $d(x_{di})$ and wild type (x_m) reads:

$$x_{di}/x_m = p^d \frac{w_{mm}}{w_{ii}} f_{di}(W) \tag{16}$$

where p is the quotient introduced in eq. (14). For each error class defined by the Hamming distance d, $w_{di} = w_{ii}/(w_{mm} - w_{ii})$ refers to an individual i of this class, while $f_{di}(W)$ is obtained through the following iteration:

$$f_{1i}(W) = W_{1i}$$

$$f_{2i}(W) = W_{2i} \left\{ 1 + \sum_{j=1}^{\binom{2}{1}} f_{1j}(W) \right\}$$

$$f_{3i}(W) = W_{3i} \left\{ 1 + \sum_{j=1}^{\binom{3}{1}} f_{1j}(W) + \sum_{j=1}^{\binom{3}{2}} f_{2j}(W) \right\} \tag{17}$$

$$f_{di}(W) = W_{di} \left\{ 1 + \sum_{j=1}^{\binom{d}{1}} f_{1j}(W) + \ldots + \sum_{j=1}^{\binom{d}{d-1}} f_{d-1,j}(W) \right\}$$

In the sum terms, the j's refer to corresponding precursors of di, i.e. in $f_{1j}(W)$ to all d one-error precursors, or in $f_{d-1,j}(W)$ to all $(d-1)$ error precursors of d_i. Note that, due to its iterative nature, f_{di} in its last term includes d factorial d-fold products of W-terms.

According to definition of w_{mm} as the largest diagonal coefficient and in line with the second-order perturbation approximation, singularities among the hyperbolic terms W_{di} are precluded, although this approximation still allows for quite large values, possibly reaching several orders of magnitude, depending on the magnitude of ν. For most mutants with $w_{mm} \gg w_{ii}$, the w_{di} values become quite small or, for non-viable mutants, even reach zero. If all mutants were of such a kind (i.e. $w_{ii} \approx 0$) the distribution of the individual relative population numbers in a mutant class d would simply resemble p^d. Expression (16) comprises all contributions of mutant states (classes $1 \ldots d$) represented in figure (3), including stepwise mutations as well as any jumps up to the length d occurring in the direction 0 to d. Contributions from reverse mutations or looped routes through states outside the

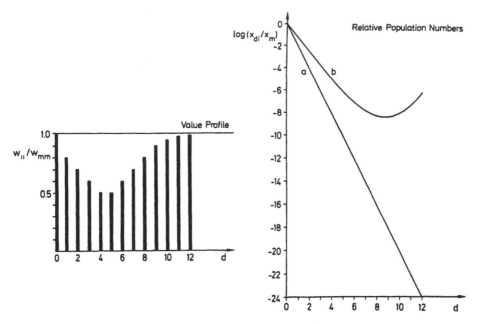

FIGURE 5 Example for the modification of the population distribution of mutants (i) as a consequence of their selective values w_{ii} being close to w_{mm}, the selective value of the wild type. At the left a particular value profile is shown for which the relative population distribution (x_{di}/x_m) was calculated according to eq. (16), assuming $p = 10^{-2}$. This distribution is reflected in curve b), while curve a) shows the distribution p^d, i.e., assuming selective values $w_{ii} \ll w_{mm}$ for the mutants. For $d = 12$, the particular "valuable" mutants appear 10^{17} times more frequently than valueless mutants.

diagram have been neglected. Their contributions are at least by a factor p smaller than terms considered in this approximation. (Note that p, depending on sequence length, usually is a very small quantity.) In the quasi-species distribution, some mutants usually have selective values w_{ii} close to w_{mm}. Accordingly, the distribution p^d may be drastically modified. This is especially true if in the total mutant spectrum certain regions such as the one represented in figure (3) (i.e. "mountain regions" in the value landscape) are involved. Experiments suggest that the value distribution is by no means random but rather clustered in such more or less connected regions of value space. As mentioned before, it somehow is similar to height distribution on the two-dimensional surface of earth, except that it refers to the ν-dimensional sequence space. In such a region where selective values are not too much different from w_{mm}, quite large modifications of population variables as compared to "low-valued" regions (i.e. "planes" in the value landscape) may occur, which—due to their multiplicative nature (produces of W-terms)—may reach large orders of magnitude. An example is presented in figure (5). Accordingly the mutant distribution in high-valued regions may reach very far into sequence space. While low-valued region may populate mutants (cf. the example discussed above) only to Hamming distances not exceeding $d = 3$ to 4, in high- valued regions mutants with $d = 10$ to 20 may well be populated. Since in a clustered value distribution the advantageous copy also is expected to appear at "mountain" rather than at "plane" sites, there is a guiding of the evolutionary process through sequence space which—due to the high dimensionality of this space—can be very efficient. The evolutionary process, so to speak, proceeds along the multiply- interconnected ridges in sequence space. Such a guiding, on the other hand, would be absent if selective values were randomly distributed in sequence space (i.e. non-clustered).

The foregoing treatment answers some of the questions about the distribution of a quasi-species in sequence space. Being deterministic in nature, such a treatment is limited to those states which reproducibly are populated to a statistically significant extent. This deterministic treatment has been complemented by J. S. McCaskill[19] so as to include a stochastic description of the rare events which may or may not happen at the periphery of the mutant population. These events, the probability of which depends on the population structure of the localized distribution in sequence space, may include destabilization of the former (meta-stable) wild type and a complete reshuffling of the mutant population. The tendency towards localization which then is steadily challenged by newly arriving mutants in peripheral regions of value mountains causes the long-range evolutionary process to be steplike. Through the preferred population of mountain site of the value landscape and due to a clustered distribution of mountain sites, the evolutionary process is tuned to proceed towards high fitness values, although it may never reach the global maximum and certainly will miss isolated peaks.

Attempts have been made on the basis of the quasi-species model to develop a cohesive theory for this diffusion-like migration through a multi- dimensional space with randomly distributed sources and sinks. W. Ebeling[20] and R. Feistel[5] et al. emphasize the equivalence of the problem to the quantum-mechanical motion of electrons in random fields. A similar equivalence to spin glasses was demonstrated by P. W. Anderson.[21,22] As was stressed in this paper, the success of these models will greatly depend on a more precise knowledge of the non-random value distribution which we think in any concrete case can be obtained only through appropriate experiments.

6. CONCLUSIONS: MACROMOLECULAR EVOLUTION IN NATURE AND IN THE LABORATORY

Knowing the regularities associated with evolutionary adaptation in systems of self-reproducing entities provokes two questions:

■ What are the constraints under which such a process could have taken place in nature?
■ Is it possible to provide conditions that allow a simulation of such processes in the laboratory?

There are, of course, many problems of chemical nature associated with early evolution. In particular, the synthesis of nucleic acid-like compounds under prebiotic conditions is a problem that involves still many unanswered questions despite the remarkable successes that have been gained in recent years.[23,24] However, this is not a problem to which this paper is intended to make any contribution. If we talk about laboratory experiments, we shall suppose the existence of biochemical machinery for the synthesis of nucleic acids. Likewise we shall assume that at some stage of evolution such machinery was ready to produce the huge variety of possible sequences we have talked about. The problem to be considered here is the constraints under which these sequences could evolve to optimal performance.

The spatial and temporal constraints of a planetary laboratory certainly cannot be matched by any man-made machine. Yet the discrepancy in orders of magnitude is not as impressive if we compare it with the orders of magnitude of possible sequences that had to be narrowed down in the evolutionary process through guided natural selection. All oceans on earth (covering an area of about 361 million square kilometers and having an average depth of 3800m) contain "only" about 10^{21} liters of water. The water content of lakes and other fresh water sources is about four orders of magnitude lower. Hence realistic numbers of macromolecules that could be tested at any instant in nature may have been as large as 10^{30} or more, but could not exceed the order of magnitude of 10^{40} (note that 10^{42} macromolecules dissolved in all oceans would produce a highly viscous broth). Moreover, the time available for macromolecular evolution up to optimal performance was smaller (and possibly much smaller) than 10^9 years $\approx 4 \times 10^{16}$ seconds. On the other hand, to produce an RNA sequence comprising a thousand nucleotides, even if a well-adapted and efficient replicase is used, requires times of a few seconds to a minute. Hence the maximum number of sequences on earth that ever could have been tested must be much below an order of magnitude of 10^{60} and may even barely reach 10^{50}.

In laboratory experiments, one is typically dealing with some 10^{12} to 10^{15} RNA or DNA sequences, viruses or microorganisms, which in large-scale projects may be extended to 10^{18} to 10^{20} entities. The time one may devote to such experiments, i.e. the time typically spent on Ph.D. work, is of the order of magnitude of 10^7 to 10^8 seconds.

These differences of spatial and temporal constraints on evolutionary processes under laboratory vs. planetary conditions have to be compared with the reduction of orders of magnitude achieved by such processes. A sequence comprising 100 nucleotides has 10^{60} different alternatives. If selection had to be achieved through random testing of these alternatives, macromolecular evolution under planetary constraints would have reached its limits with such relatively short sequences. We have good reasons to assume that limitations of this kind became effective only at appreciably larger length. Let me quote two reasons:

1. RNA viruses, which disseminate and adapt to environmental constraints on the basis of straightforward replication only (i.e., using enzymes without sophisticated error correction) reach typical lengths of 10^3 to 10^4 nucleotides. The lengths of genes usually are around 10^3 nucleotides. Domain structures of proteins suggest that larger lengths may have been achieved only through gene doubling or fusion. Hence one may assume that gene elongation in evolution on the basis of straightforward replication using copying errors as the source of adaptation could proceed to gene lengths of around 10^3 nucleotides. The number of alternatives here is 10^{600}.

2. Q_β-replicase has been found to be able to produce de novo RNA sequences that can be adapted to strange environmental conditions.[25,26,27,28] In such experiments, for instance, ribonuclease T_1 resistant sequences have been obtained which in the presence of normally "lethal" doses of this enzyme grow as efficiently as optimal wild types do under normal conditions. Ribonuclease T_1 cleaves RNA sequences at exposed, unpaired G-residues. Resistance to cleavage therefore requires the sequence to refold in such a way that all exposed G-residues become inaccessible to the enzyme, e.g. through base pairing or hiding inside the tertiary structure. The minute fraction of sequences present in these experiments (cf. above) obviously was sufficient to allow for adaptation to optimal performance within a relatively small number of generations. (The bulk of sequences of this particular size would include some 10^{120} different mutants.)

Evolutionary adaptations, as was shown in this paper, is equivalent to hill climbing along proper ridges in the rugged value landscape, rather than to an unbiassed random walk. The total number of alternative mutants therefore is not as important as the existence of advantageous mutants within reachable distances. Proper refolding may always be possible if a sufficient number of residues are exchanged. The examples in this paper suggest that for sequence lengths of 300 nucleotides mutation distances of 10 to 20 nucleotides along routes of selectively advantageous mutants can be bridged under typical laboratory conditions, possibly sufficient for reshaping the phenotype into a more advantageous conformation.

Such conjectures, of course, have to be tested experimentally. We have embarked on such tests following essentially two routes:

1. If the value of the phenotype is properly represented by the selective value of the genotype (which essentially rates efficiency of reproduction) serial transfer under constrained growth conditions, as first applied by S. Spiegelman and coworkers,[29] may be an efficient tool to scan through large mutant

populations. The theory suggests optimal conditions for such experiments, which includes the growth conditions (usually exponential), the dilution factor (variable and usually as small as possible in order to keep the mutant population close to stationary conditions) and the mutation rate (which is to be reproducibly regulated around the error threshold allowing for some type of annealing). The required speed and control in those experiments suggests automation.

2. The more interesting case is that of independent evaluation of phenotypes and their evolutive adaptation to various tasks. In this case natural selection is to be replaced by artificial selection, while reproducibility, controlled mutability and amplificability of phenotypes still requires their genotypic representation. Hence, mutated sets of their genotypes have to be cloned and

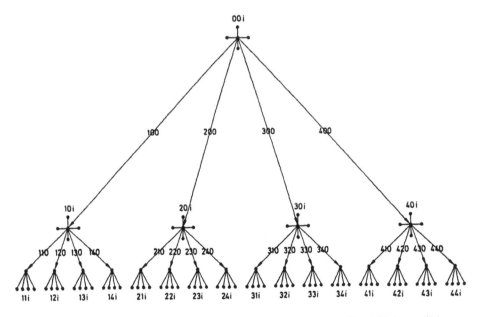

FIGURE 6 Production of hierarchically ordered mutant spectra. The initial "seed" is a wild type that is reproduced with a specified (high)error rate over a few generations, yielding mutants with a large average Hamming distance. Those mutants are cloned and the procedure is repeated with a reduced error rate. Iteration leads finally to addressable clones that can be classified according to their kinship distances. Topographic maps displaying the relation between phenotypic properties andthe known Hamming distances between clones then can be constructed. The procedure involves large-scale cloning and parallel assessment of phenotypic properties. The cloning procedure requires reproducibly controlled error rates and parallel serial dilution.

screened for advantageous phenotypic properties. Since the number of clones that can be handled by any automated device is limited (note that samples in the serial transfer technique may easily include 10^{12} sequences), the search for advantageous mutants must be correspondingly systematized. Natural selection is efficient because of the large number of mutants, among which the most advantageous ones are preferably populated. The connectivity of the value distribution then provides for guidance to the optimum. This principle may be utilized by special cloning devices. For this purpose it is necessary to produce hierarchically ordered mutant spectra with known (average) mutual distances (cf. figure 6). After screening for particular functional values, the known interclonal mutation distances then suffice for reconstructing the value landscape and for identifying the "mountainous" regions. Mutants belonging to corresponding clones are used in the next generation to scrutinize the procedure. The target structure of the phenotype and its genotype resp. then is to be reached through multiple iterations. The procedure is like representing mountains on maps. For drawing the map, it is sufficient to use a relatively small number of coloured dots where the colour refers to the height of the corresponding point in nature. What is necessary is that the chosen points are sufficiently distributed and that their relative mutual position are known. A non-random distribution of heights such as for landscapes on earth can be easily localized this way, through iteration finally with any degree of resolution. The essential feature of this technique is the production of hierarchical mutant spectra with known (average) distances. This requires the ability to reproducibly vary mutation rates, so that the error threshold is violated in a controlled way. This corresponds to some kind of annealing simulation which is utilized for optimization in multiple variable systems.[30] An automated machine for production, synchronous incubation and parallel screening of a large number of such hierarchically ordered clones is under construction.

REFERENCES

1. M. O. Dayhoff, *Atlas of Protein Sequence and Structure*, (Washington, D.C.: Nat. Biomed. Res. Found., 1972), Vol. **5**, D7ff.
2. J. J. Birktoft and D. M. Blow, *J. Mol. Biol.* **68**, 187 (1972).
3. J. Drenth, W. G. J. Hol, J. N. Jansonius and R. R. Koekoek, *Cold Spring Harbor Symp. Quant. Biol* 36, 107 (1971).
4. I. Rechenberg, *"Evolutionsstrategie" Problemata* (Frommann-Holzboog, Stuttgart-Bad Canstatt, 1973).
5. R. Feistel and W. Eberling, *BioSystems* 15, 291 (1982).
6. M. Eigen and P. Schuster, *Naturwissenschaften* **64**, 541 (1977).
7. M. Eigen, *Naturwissenschaften* **58**, 465 (1971).

8. B. L. Jones, R. H. Enns and S. S. Rangnekar, *Bull. Math. Biol.* **38**, 15 (1976)
9. C. J. Thompson and J. L. McBride, *Math. Biosci.* **21**, 127 (1974).
10. J. .S. McCaskill, *J. Chem. Phys.* **80** (10), 5194 (1984).
11. E. Domingo, R. A. Flavell and C. Weissmann, *Gene* **1**, 3 (1976); E. Domingo, M. Davilla and J. Ortin. *Gene* **11**, 333 (1980).
12. E. Domingo, D. Sabo, T. Taniguchi and C. Weissmann, *Cell* **13**, 735 (1978).
13. Philipp Mekler, Inaugural Dissertation, Universität Zürich, 1981.
14. Ch. Biebricher, M. Eigen and W. C. Gardiner, Jr., *Biochemistry* (1985), in press.
15. C. K. Biebricker, M. Eigen and W. C. Gardiner, *Biochemistry* **22**, 2544 (1983).
16. C. K. Biebricher, M. Eigen and W. C. Gardiner, Jr., *Biochemistry* **23**, 3186 (1984).
17. R. Saffhill, H. Schneider-Bernloehr, L. E. Orgel and S. Spiegelman, *J. Mol. Biol.* **51**, 531 (1970).
18. F. R. Karmer, D. R. Mills, P. E. Cole, T. Nishihara and S. Spiegelman, *J. Mol. Biol.* **89**, 719 (1974).
19. J. S. McCaskill, *Bio. Cybernetics* **50**, 63 (1984).
20. W. Ebeling, A. Engel, B. Esser and R. Feister, *J. Statist. Phys.* **37**, (314), 369 (1984).
21. P. W. Anderson, *Proc. Natl. Acad. Sci. USA* **80**, 3386 (1983).
22. D. L. Stern and P. W. Anderson, *Proc. Natl. Acad. Sci. USA* **81**, 1751 (1984).
23. R. Lohrmann, P. K. Bridson and L. E. Orgel, *Science* **208**, 1464 (1980).
24. G. F. Joyce, G. M. Visser, C. A. A. van Boeckel, J. H. van Boom, L. E. Orgel, and Y. van Mestrenen, *Nature* **310**, 602 (1984).
25. M. Sumper and R. Luce, *Proc. Natl. Acad. Sci. USA* **72**, 162 (1975).
26. C. K. Biebricher, M. Eigen and R. Luce, *J. Mol. Biol.* **148**, 369 (1981).
27. C. K. Biebricher, M. Eigen and R. Luce, *J. Mol. Biol.* **148**, 391 (1981).
28. C. K. Biecricher in *Evolutionary Biology*, Vol. **16**, 1. Ed. by M. K. Hechet, B. Wallace and C. T. Prance (Plenum, New York, 1983).
29. D. R. Mills, R. L. Peterson and S. Spiegelman, *Proc. Natl. Acad. Sci.* **58**, 217 (1967).
30. S. Kirkpatrick, C. D. Gelatt, Jr. and H. P. Vecchi, *Science* **220**, 671 (1983).

MARCUS W. FELDMAN
Department of Biological Sciences, Stanford University, Stanford, CA 94305

Evolutionary Theory of Genotypes and Phenotypes:
Towards a Mathematical Synthesis[1]

INTRODUCTION

The structures and functions of an organism that can be observed and measured are called its *phenotype*. Some parts of the phenotype, e.g., blood groups or enzyme concentration, require more sophisticated calibration than is amenable to direct observation. Nevertheless, they are in principle observable and are therefore phenotypes. The *genotype*, on the other hand, is defined entirely by the sequence of nucleotides that make up the DNA. For a given genotype, different phenotypes may be realized, depending on the environment in which the organism finds itself. The *norm of reaction* of a genotype is the pattern of the phenotypes that can be realized by placing that genotype in some range of environments.

The variation that Darwin perceived was phenotypic; evolution was the process of the conversion of phenotypic variation between individuals into phenotypic variation between populations and species. The transmission of this variation from parent to child was assumed by Darwin and Galton to be blending in character: the expected phenotype of a child was the average of its parents' phenotypes. This produced the paradox that phenotypic variation should eventually disappear, and

[1] The writing of this paper supported in part by NIH grants GM 28016 and 10452.

it was not until the rediscovery of Mendel's particulate theory of transmission that the paradox was resolved. Mendel's phenotypic differences were the result of simple genotypic differences whose transmission could be described quite precisely. Under Mendelian transmission, Hardy and Weinberg were able to show that phenotypic variation, resulting from genetic differences of the Mendelian kind, is conserved. Insofar as the genotype contributes to the phenotype (as described by the norm reaction), natural selection on the phenotype, acting via the environment, results in the conversion of genotypic differences between individuals into genotypic variation between populations and species.

Fisher (1918) was the first to demonstrate mathematically how Mendelian qualitative differences could be translated into metrical or quantitative variation. His theory allowed quantification of expected statistical relationships between the phenotypes of relatives. It was not, however, an evolutionary theory, and did not allow for the action of natural selection on the phenotype. Nevertheless, animal breeders subsequently used Fisher's theory in attempting to predict the genetic consequences of artificial selection on the pheonotype (see, for example, Lewontin, 1974, p. 15).

The serious mathematical difficulty inherent in the construction of a formal theory of phenotypic change was recognized early in the history of population genetics. The mathematical theory originated by Fisher, Wright and Haldane was genotypic in nature, and in their mathematical models, phenotypic differences were identified with genotypic differences at one or a very small number of genes. In these models, natural selection acted on the genotype and evolution occurred as genotypic frequencies changed. The body of mathematical evolutionary theory developed over the past 65 years has addressed genotypic evolution, although, as will be mentioned later, there have been a few attempts to mathematically model phenotypic change under natural selection. I will first introduce the standard mathematical formulation of genotypic natural selection and indicate the consequences of departures from this standard paradigm. I will then introduce a formulation that allows some degree of synthesis of phenotypic and genotypic evolution and comment on its potential relevance to the evolution of behavior.

SELECTION DUE TO VIABILITY DIFFERENCES AT A SINGLE GENE

The theory of selection on the genotype has been most extensively studied under the assumption that the selection is due to differences among genotypes in their ability to survive from birth (fertilized egg) to adulthood. We call this *viability* selection. In the case of a single gene, A, suppose that the alternative forms of it, its *alleles*, are $A_1, A_2, \ldots, A_r$ and that in a given generation the fraction of A_i is x_i, $i = 1, 2, \ldots, r$ with $\sum_i x_i = 1$. The genotypes are specified by the pairs $A_i A_j (i, j = 1, 2, \ldots, r)$ with $A_i A_i$ called homozygotes and $A_i A_j$, heterozygotes. Now suppose that the generations are nonoverlapping and that the relative survival probability of $A_i A_j$

is w_{ij}. The matrix $\| w_{ij} \| = \underline{W}$ is called the *viability* matrix. The frequencies x_i' in the next generation of adults, after selection, are specified by

$$x_i' = \frac{x_i \sum_{j=1}^{r} w_{ij} x_j}{\sum_{i=1}^{r} \sum_{j=1}^{r} w_{ij} x_i x_j}. \tag{1}$$

The denominator of (1), $W(\underline{x}) = \sum_i \sum_j w_{ij} x_i x_j$, is the mean fitness of the population with frequency vector $\underline{x} = (x_1, x_2, \ldots, x_r)$. For brevity, denote the transformation (1) by

$$\underline{x}' = \underline{T}(\underline{x}). \tag{1}$$

I now list some of the best known properties of the transformation (1), each of which should be regarded as a *qualitative* statement about evolution at a single locus.

PROPERTY (I) . The Fundamental Theorem of Natural Selection originally enunciated by Fisher (1930) can be stated as follows

$$W\left(\underline{T}(\underline{x})\right) \geq W(\underline{x}) \tag{2}$$

with equality holding if and only if $\underline{x} = T\underline{x}$, that is at equilibria (or fixed points) of (1). Qualitatively, this is an elegant formal expression of the Darwinian idea that the mean fitness of a population should increase over time. More recent "strategy" terminology would have it that a stable equilibrium of (1) is an optimum for the population, since it (locally) maximizes the mean fitness. The most elegant proof of (2) is due to Kingman (1961a).

PROPERTY 2. A *polymorphism* is a fixed point of (1) at which more than one allele has positive frequency. A *complete polymorphism* is a polymorphism with $x_i > 0$ for all $i = 1, 2, \ldots, r$. There exists at most one complete polymorphism, and it is globally stable if and only if the matrix $\underline{W}$ has one positive and $r - 1$ negative eigenvalues (Kingman, 1961b). Note that there can be two stable polymorphisms, for example, one with A_1 and A_2 and another with A_3 and A_4, but stability of the complete polymorphism precludes the stability of any other equilibrium of (1).

PROPERTY 3. When w_{ij} $(i, j = 1, 2, \ldots, r)$ are chosen randomly from a uniform distribution of [0,1], the probability that the matrix $\underline{W}$ allows a stable complete polymorphism decreases as a function of r according to

$$exp \left[-\frac{r^2}{2} \log r \right]$$

(Karlin, 1981). In other words, for $r \geq 5$ if viabilities were assigned randomly by the environment, complete polymorphism would be highly unlikely.

PROPERTY 4 . When $r = 2$ heterozygote advantage, that is $w_{12} > w_{11}$, w_{22} is sufficient for a stable polymorphism with A_1 and A_2. In general, however, $w_{ij} > w_{ii}$, w_{jj} (all i, j) is not sufficient for a complete polymorphism to be stable. The following is a simple 4-allele counterexample

$$\underline{W} = \begin{bmatrix} (1-s)^2 & 1-s & 1-s & 1 \\ 1-s & (1-s)^2 & 1 & 1-s \\ 1-s & 1 & (1-s)^2 & 1-s \\ 1 & 1-s & 1-s & (1-s)^2 \end{bmatrix},$$

with $s < 1$.

COMPLICATIONS DUE TO MORE THAN ONE GENE

Consider two genes A and B such that the alleles at A are A_1 and A_2, while B_1 and B_2 are those at the second. For the present purposes, it will be sufficient to consider just two alleles at each locus. The gametic types that may exist in the population are then $A_1 B_1$, $A_1 B_2$, $A_2 B_1$ and $A_2 B_2$. There are ten distinct genotypes that can be constructed from these $A_1 B_1 / A_1 B_1$, $A_1 B_1 / A_1 B_2$, etc., but the two double-heterozygotes $A_1 B_1 / A_2 B_2$, $A_1 B_2 / A_1 B_2$ have the same viability. This is based on the assumption that their gene products must be the same so that they cannot be distinguished by any environmental stress. In principle, A and B could be located on different chromosomes, and therefore inherited independently, both in Mendelian fashion. The genotype $A_1 B_1 / A_2 B_2$ may produce all four gametes $A_1 B_1$, $A_1 B_2$, $A_2 B_1$ and $A_2 B_2$. If the genes are on the same chromosome, they may be *linked*. If only the parental gametes $A_1 B_1$ and $A_2 B_2$ are produced by $A_1 B_1 / A_2 B_2$, the loci are said to be *absolutely linked*. If a fraction c of $A_1 B_2$ and $A_2 B_1$ is produced, then c is called the *recombination fraction*, and the larger is c, the looser is the linkage between A and B. The biology of recombination precludes c from being greater than 0.5.

Now let x_1, x_2, x_3, and x_4 be the frequencies of $A_1 B_1$, $A_1 B_2$, $A_2 B_1$ and $A_2 B_2$, respectively with $P_{A_1} = x_1 + x_2$ and $P_{B_1} = x_1 + x_3$, the allele frequencies of A_1 and B_1, respectively. Suppose that the relative viability of genotype i, j $(i, j =$

$1, 2, 3, 4)$ be v_{ij}. Then the following transformation produces the gamete frequencies x'_1, x'_2, x'_3, x'_4 in the next generation

$$V\, x'_1 = x_1 v_1. - cD_1 v_4 \qquad (3a)$$
$$V\, x'_2 = x_2 v_2. + cD\, v_{14} \qquad (3b)$$
$$V\, x'_3 = x_3 v_3. + cD\, v_{14} \qquad (3c)$$
$$V\, x'_4 = x_4 v_4. - cD\, v_{14}, \qquad (3d)$$

where $v_{14} = v_{23}$ is the viability of the double heterozygotes, $v_1. = \sum_{j=1}^{4} v_{ij} x_j$, $V = V(\underline{x}) = \sum_{i=1}^{4} x_i v_i.$, and $D = x_1 x_4 - x_2 x_3$. The mean viability V has a similar meaning to W in (1). D is called the linkage disequilibrium, and because we may write

$$x_1 = P_{A_1} P_{B_1} + D \qquad (4a)$$
$$x_2 = P_{A_1} P_{B_2} + D \qquad (4b)$$
$$x_3 = P_{A_2} P_{B_1} + D \qquad (4c)$$
$$x_4 = P_{A_2} P_{B_2} + D \qquad (4d)$$

it is clear that D measures the departure of the gamete frequencies from being constructed solely from allele frequencies.

The evolution of the linked pair of genes A and B is described by the trajectory of the recursion system (3) upon iteration. It is not my purpose to survey all that is known about the properties of (4). Suffice it to say that a complete accounting of the fixed point and their stability is not available for more than a few special forms of the matrix $\underline{V} = \| v_{ij} \|$. These cases are surveyed in Ewens (1979, ch. 6). I wish here to point out several qualitative differences from the simple one-locus theory.

1. The mean fitness does not usually increase. This was first noted by Moran (1964) and makes the analysis of (3) technically difficult because there is no natural Lyapounov function. For c very small, the system (3) is close to the one-locus, 4-allele model for which mean fitness does increase. But when the linkage is loose (recombination close to 0.5), it is not clear what measure of viability is "optimized." The significance of this for strategy reasoning has recently been analysed by Eshel and Feldman (1984).

2. For $c > 0$, there may exist more than one complete polymorphism. For all of the special cases of $\underline{V} = \| v_{ij} \|$ for which solutions exist, there is a complete polymorphism with $D = 0$. This has been of special interest because it reflects a lack of interaction between the genes in the consequences of selection. Franklin and Feldman (1977) and Karlin and Feldman (1978) have produced cases in which a stable fixed point with $D = 0$ and one with $D \neq 0$ exist simultaneously. The maximum number of stable complete polymorphisms that

can coexist has not exceeded 4 in any fully analysed case. Unlike the one-locus case, complete polymorphism and incomplete polymorphism may be simultaneously stable (for example, Feldman and Liberman (1979)).

3. There is a recent numerical finding (Hastings, 1981) that for some choices of $\underline{V} = \| v_{ij} \|$, stable cycles result from the iteration of (3). Of course this is impossible with one locus.

SELECTION DUE TO FERTILITY DIFFERENCES

In standard demographic practice, the number of offspring is usually measured per female. That is, the theoretical framework is essentially unisexual. In fact, the number of offspring should be considered *per mating*. In genetic terms, fertility is a property of both male and female parental genotypes. Consider a single gene with alleles $A_1, A_2, \ldots, A_r$ such that the frequencies of genotypes $A_i A_j$ just prior to mating are $X_{ij}(i, j = 1, 2, \ldots, r)$. The relative fertility of the mating between $A_i A_j$ and $A_m A_n$ is f_{ijmn}. (Without loss of generality, the order of the sexes can be ignored here; see, for example, Feldman et al., 1983.) Then, the frequencies of the genotypes just prior to mating in the next generation, after the fertility selection and Mendelian segregation have acted, are (see, for example, Ewens, 1979)

$$
\begin{aligned}
F X'_{ii} = & f_{iiii} X_{ii}^2 + \frac{1}{2} \sum_{m \neq 1} f_{iiim} X_{ii} X_{im} + \frac{1}{2} \sum_{j \neq i} f_{ijii} x_{ij} x_{ii} + \\
& \frac{1}{4} \sum_{j \neq i} \sum_{m \neq i} f_{ijim} X_{ij} X_{im}
\end{aligned}
\tag{5a}
$$

for the homozygote $A_i A_i$, and

$$
\begin{aligned}
F X'_{ij} = & (f_{iijj} + f_{jjii}) X_{ii} X_{jj} + \frac{1}{2} \sum_{m \neq j} f_{iijm} X_{ii} X_{jm} + \\
& \frac{1}{2} \sum_{m \neq i} f_{imjj} X_{im} X_{jj} + \frac{1}{4} \sum_{m \neq i} \sum_{n \neq j} f_{imjn} X_{im} X_{jn}
\end{aligned}
\tag{5b}
$$

for the heterozygotes. Here F is the normalizer chosen so that $\sum \sum X'_{ij} = 1$. It is in fact the "mean fertility" of a population whose genotypic frequency array is $\underline{X} = (x_{11}, x_{12}, \ldots, x_{rr})$.

Little is known about the properties of the recursion (5) in general. For some special cases, however, a complete evaluation of the dynamic has been possible. The summary that follows is far from complete, but indicates the striking differences from expectations under viability selection.

1. Even with a single gene, the mean frequency does not necessarily increase. Specific examples were provided by Pollack (1978).
2. It is not necessary that (5) admits any stable fixed points. This has been shown by Hadeler and Liberman (1975).
3. Even with two alleles it is possible for both monomorphism and complete polymorphism to be stable, and the maximum number of admissible polymorphic equilibria is not known (Bodmer, 1965; Hadeler and Liberman, 1975; Feldman et al., 1983).
4. With two loci and two alleles at each, some progress has been made when the fertility depends only on the number of heterozygotes counted in both parents. (This can go from zero to four under these conditions.) It becomes clear that a very complicated pattern of simultaneous stability of many complete (and incomplete) polymorphisms is possible (Feldman and Liberman, 1985). It is highly unlikely that any straightforward theory of optimization will emerge when fitness is measured at the fertility level.

The importance of these remarks to evolutionary biology must be viewed in the context of empirical knowledge about fitness. As summarized by Feldman and Liberman (1985), experimental evidence is overwhelming that fertility differences contribute far more significantly to net fitness variation than do viability effects. This suggests that the simple and elegant theory of viability selection needs reassessment as to its relevance for general evolutionary theory.

The primary reason for the great increase in complexity that occurs with fertility selection is that *gene frequencies* are not sufficient to specify the evolutionary dynamic. Genotype frequencies are required, and there are many more of these. The same increase in complexity occurs in the study of models with mixed mating systems in which individuals inbreed or outbreed with specific probabilities.

SIMULTANEOUS EVOLUTION OF PHENOTYPE AND GENOTYPE

Natural selection acts via the environment on the phenotype. Unfortunately, rules of transmission for the phenotype are not as simple as Mendel's rules for genetic transmission. Fisher's theory identifies the transmission of genes and phenotypes by taking each genotype to contribute in a precise way to the phenotype. Attempts to impose natural selection on this continuous variation have succeeded only under special assumptions. The main assumptions required to produce tractable analysis are that the phenotype has a Gaussian distribution and that the form of the natural selection is Gaussian. The latter entails that individuals close to some optimum survive better than those far from it according to the normal density function. Even under these conditions, the evolutionary studies by Kimura (1965), Cavalli-Sforza and Feldman (1976), Lande (1976) and Karlin (1978) incorporate genotypic transmission in ways that are not easy to relate to the action of single Mendelian genes under viability selection.

Cavalli-Sforza and I have taken a simpler approach. We start with a dichotomous phenotype taking the value 1 and 2, and a single gene with alleles A and a. There are then six phenogenotypes: AA_1, AA_2, Aa_1, Aa_2, aa_1, aa_2. Natural selection acts only on the phenogenotype so that the relative fitnesses of phenotype 1 and 2 are 1 and $1 - s$, respectively. Mendelian transmission governs the gene, but the phenotype must be transmitted in a more complex way.

Suppose that the parents' phenotypes and genotypes, and the offspring's genotype influences the probability that the offspring is of phenotype 1. Then we may represent the transmission process with a set of sixty parameters $\beta_{ijk,\ell m}$, where i, j, k represent the mother's, father's and offspring's genotype, respectively, and ℓ and m are the mother's and father's phenotype. Thus, $i, j, k = 1, 2, 3$ for AA, Aa, aa, respectively with $\ell, m = 1, 2$. Using these parameters for *parent-to-offspring* transmission and the selection coefficient s, a recursion system for the six phenogenotypes may be developed. Felman and Cavalli-Sforza (1976) analysed the case where the dependence on i, j and, say, m was removed; the probability of an offspring being phenotype 1 was a function of its own genotype and of the phenotype of one parent only, the "transmitting parent." Although we envisage the general formulation to apply to the transmission of some learned behavior, the quantitative model is equally relevant to the vertical transmission of an infectious disease where infectiousness of the transmitter and susceptibility to infection are genetically influenced.

Among the more interesting findings to emerge from the co-evolutionary analysis are:

1. The average fitness does not always increase throughout the evolutionary trajectory, although it appears to do so locally in the neighborhood of stable equilibria.

2. Heterozygote advantage in transmission of an advantageous trait does not guarantee a polymorphism. Thus, if Aa transmits (or receives) phenotype 1 better than AA and aa, then even if $s > 0$, there may not be a stable polymorphism.

3. When there is no selection ($s = 0$) on the phenotype, the phenotype frequencies change only under the influence of transmission (the β's). The rate of evolution for this is orders of magnitude faster than if $s \neq 0$. The reason for this is in the structure of the recursions. With $s = 0$, the recursions are quadratic. With $s \neq 0$, they are ratios of quadratics. These ratios arise in population genetics and are relatively slow moving.

In general, when there are fitness differences among the phenotypes that are described by a vector $\underline{s}$, and a transmission rule that is represented by a vector $\underline{\beta}$, the evolution of the vector $\underline{x}$ of phenogenotype frequencies may be written

$$\underline{x}^{t+1} = T(\underline{s}, \underline{\beta}; \underline{x}^t).$$

The analytic approach may then address such issues in behavioral ecology as the following: *Incest Taboo*: if the phenotypic dichotomy is to breed with a relative or not, do genes which favor outbreeding win in the evolutionary race (Feldman and Christiansen, 1984)? *Evolution of learning*: if one set of genes entails that the phenotypes of their carriers are entirely genetically determined, while another set of genes allows these phenotypes to be acquired in a non-genetic manner, which genes will succeed? *Evolution of altruism*: if the phenotypic dichotomy is to perform or not to perform altruistic acts (that is, to sacrifice one's own fitness so that other individuals, for example relatives, might increase their fitness), does the evolutionary dynamic depend on whether the behavior is innate or learned?

The design of appropriate parameter sets $\underline{\beta}$ for such studies requires the synthesis of genetic and social science thinking. I believe that for behavioral evolution to progress beyond the most rudimentary of genetic approaches, a synthetic yet rigorous approach of the kind outlined above should be pursued.

REFERENCES

Bodmer, W. F. (1965), "Differential Fertility in Population Genetics Models," *Genetics* **51**, 411–424.

Cavalli-Sforza, L. L. and M. W. Feldman (1976), "Evolution of Continuous Variation: Direct Approach through Joint Distribution of Genotypes and Phenotypes," *Proc. Natl. Acad. Sci. USA* **73**, 1689–1692.

Eshel, I. and M. W. Feldman (1984), "Initial Increase of New Mutants and Some · Continuity Properties of ESS in Two-locus Systems," *Amer. Natur.* 124, 631–640.

Ewens, W. J. (1979), *Mathematical Population Genetics* (Berlin: Springer Verlay).

Feldman, M.W. and L. L. Cavalli-Sforza (1976), "Cultural and Biological Evolutionary Process; Selection for a Trait under Complex Transmission," *Theor. Pop. Biol* **9**, 238–259.

Felman, M. W., F. B. Christiansen (1984), "Population Genetic Theory of the Cost of Inbreeding," *Amer. Natur.* **123**, 642–653.

Felman, M. W., F. B. Christiansen and U. Liberman (1983), "On Some Models of Fertility Selection," *Genetics* **105**, 1003–1010.

Felman, M. W.and U. Liberman (1979), "On the Number of Stable Equilibria and the Simultaneous Stability of Fixation and Polymorphism in Two-locus Models," *Genetics* **92**, 1355–1360.

Felman, M. W.and U. Liberman (1985), "A Symmetric Two-locus Fertility Model," *Genetics* **109**, 229–253.

Fisher, R. A. (1918), "The Correlation between Relatives on the Supposition of Mendelian Inheritance," *Trans. Roy. Soc. Edinburgh* **52**, 399–433.

Fisher, R. A. (1930), *The Genetical Theory of Natural Selection* (Oxford: Clarendon Press).

Franklin, I. R. and M. W. Feldman (1977), "Two Loci with Two Alleles: Linkage Equilibrium and Linkage Disequilibrium Can Be Simultaneously Stable," *Theor. Pop. Biol.* **12**, 95–113.

Hadeler, K. P. and U. Liberman (1975), "Selection Models with Fertility Differences," *J. Math. Biol.* **2**, 19–32.

Hastings, A. (1981), "Stable Cycling in Discrete-time Genetic Models," *Proc. Natl. Acad. Sci. USA* **78**, 7224–7225.

Karlin, S. (1981), "Some Natural Viability Systems for a Multiallelic Locus: a Theoretical Study," *Genetics* **97**, 457–473.

Karline, S. and M. W. Feldman (1978), "Simultaneous Stability of $D = 0$ and $D \neq 0$ for Multiplicative Viabilities of Two Loci," *Genetics* **90**, 813–825.

Kimura, M. (1965), "A Stochastic Model Concerning the Maintenance of Genetic Variability in Qualitative Characters," *Proc. Natl. Acad. Sci. USA* **54**, 731–736.

Kingman, J. F. C. (1961a), "A Matrix Inequality," *Quart. J. Math* **12**, 78–80.

Kingman, J. F. C. (1961b), "A Mathematical Problem in Population Genetics," *Proc. Camb. Phil. Soc.* **57**, 574–582.

Lande, R. (1976), "The Maintenance of Genetic Variability by Mutation in a Polygenic Character with Linked Loci," *Genet. Res. Camb.* **26**, 221–235.

Lewontin, R. C. (1974), *The Genetic Basis of Evolutionary Change* (New York: Columbia University Press).

Moran, P. A. P. (1964), "On the Nonexistence of Adaptive Topographies," *Ann. Human Genet.* **27**, 383–393.

Pollack, E. (1978), "With Selection for Fecundity the Mean Fitness Does Not Necessarily Increase," *Genetics* **90**, 383–389.

IRVEN DEVORE
Harvard University

Prospects for a Synthesis in the Human Behavioral Sciences

In the following pages I sketch my own experiences working in and between the social and biological sciences, and then present: a personal view of the present state of theory in the social sciences; the challenge to social science theory from the new paradigm in behavioral biology ("sociobiology"); some reasons why most social scientists have strongly resisted any rapprochement with evolutionary biology; and close with a few thoughts on the prospects for a unified theory of behavior.

A PERSONAL ODYSSEY

Since the views I present here will be necessarily brief and highly idiosyncractic, I first offer a summary of my experience in the behavioral sciences. My graduate work in social anthropology was at the University of Chicago, where the faculty passed on to me a largely intact version of the structural-functional paradigm, as they had received it from Durkheim via Radcliffe-Brown and Malinowski. Although I had previously had no interest in physical anthropology, I was persuaded by the remarkable Sherwood L. Washburn to undertake a field study of the social behavior of savanna baboons in Kenya. His reasoning was that highly social primates such as baboons had very complex social behavior, and that the traditional training of primatologists, in physical anthropology, comparative psychology, or mammalogy, was insufficient or inappropriate for understanding the complexity. My thesis, on the "Social Organization of Baboon Troops," was a model of the structural-functional approach, and embodied all of the implicit "group selection" presuppositions of that field. By 1964, I had begun a 12-year study of the !Kung San ("Bushmen") of the Kalahari Desert. Some 30 students and colleagues investigated a wide spectrum of topics, ranging from archaeology, demography, and nutrition to infant development, social organization, and belief systems (e.g., Lee and DeVore, 1976; Shostak, 1981). At Harvard, I had an appointment in the Department of Social Relations, taught its introductory course, and served as Chairman of its Social Anthropology "Wing." During the next 20 years, my time and that of my students has been equally divided between primate studies and hunter-gatherer studies.

In 1980, we began a similarly intensive, long-term series of coordinated studies on the Efe pygmies and Lese horticultural populations of the eastern Ituri Forest of Zaire. Our methods and research goals, however, are now very different from those we employed in the original study of the Kalahari San. Today, I am Chairman of the Anthropology Department, with a joint appointment in Biology, and consider myself a "behavioral biologist."

THEORY IN BIOLOGICAL AND SOCIAL SCIENCE

From the above it is clear that, throughout my professional life, I have vacillated between the (usually exclusive) domains of biology and social anthropology. My continuing research program has been to gather data on both primates and hunter-gatherers in an effort to better understand the evolution of human behavior. I have never been interested in social science theory *per se*; my interest has been purely pragmatic—while guiding research in primate and human studies, I have tried to remain alert to the most promising methodology and theory for such studies, no matter what the source.

By far the most important intellectual advance during my professional life has been the development of exciting new theory in vertebrate behavioral ecology, or "sociobiology." This family of theoretical advances is truly a revolution in our understanding of how evolution has shaped animal behavior. At the heart of this revolution has been the demonstration that natural selection is most accurately viewed from the "point of view" of the individual and the gene, rather than as a process that is operating at the level of the group or species. We can now, with some rigor, analyze such complex behaviors as aggression, altruism, parental care, mate choice, and foraging patterns (e.g., Dawkins, 1982; Krebs and Davies, 1984; Rubenstein and Wrangham, 1986; Trivers, 1985; and Tooby, this volume). Many of us felt, almost from the beginning, that this powerful new body of theory would also quickly revolutionize the study of human behavior. This has not proven true. To understand why a synthesis between vertebrate and human behavioral biology has been so slow to develop, I will present a very brief and opinionated view of the social science theory in which I was trained and began my early work.

To my mind, there is, at present, no deep, elegant, or even intellectually satisfying theory in social science. We continue to pay obeisance to Freud, Marx, Weber and Durkheim, but muddle through with competing and highly eclectic theories of the "middle range." This lack of fundamental theory goes to the very heart of the problems in the social sciences; I believe they have been seriously floundering for more than a decade, and are today in a state of disarray.

To illustrate the basic problem I will concentrate on the fields I know best, sociology and social anthropology. Both of these subjects can be traced back to Emile Durkheim, and then forward through the scholars I have already mentioned to contemporaries such as Talcott Parsons, Robert Merton and Levi-Strauss. Although social anthropologists have today splintered into competing factions as disparate as "ecological anthropology" and "symbolic anthropology," the underlying presuppositions on which all of these factions rest continues to be the structural-functional paradigm as originally set out by Durkheim. Because these fields analyze behavior at the group level, the mode and level of analysis are addressed to *social* phenomena: one assumes the integrity of the social group and then looks within it to analyze the roles and statuses of the group members, the enculturation of the young into group membership, etc. The social group, like a corporation, is at least potentially immortal, with individuals performing functional roles within it. The social unit, its structure and organization, is the reality (in extreme form "culture" itself becomes the reality [White, 1949]), and the individual humans are actors in the system, actively working to support its existence.

It is instructive to go back to Durkheim's original formulation (1895; transl. 1938). His metaphor for the social system was as follows: society is compared to a whole organism; within it the social institutions are like the major organs of the body (kidney, liver, etc.); individuals in the society are comparable to the cells of the organism. A healthy, functioning society is, thus, like a healthy body in which the cells are cooperating to keep the organs functional, and so on.

It is ironic, in terms of the conclusions of this chapter, that the founder of sociology originally turned to biology for his fundamental metaphor. From the modern perspective, it is easy to see the deep flaw in Durkheim's analyogy: unlike the cells in the body, individuals in a society are not perfectly genetically related, and could not be unless they were all clones of a single individual (in which case we would, indeed, expect to find a very high level of interindividual cooperation). This harmonious view of society could hardly be at greater odds with the views of sociobiology, namely, that individuals in any social group, human or otherwise, are acting out of essentially "selfish" motives (when these motives are understood, at the evolutionary level, to include behavior such as altruism toward kin, etc.)—forming coalitions, striking contracts, and gathering into self-interest groups consistent with these ultimately "selfish" motives.

Many theorists in the social sciences will be appalled to read that I am here presenting the original Durkheimian metaphor, but I am convinced that it was, in fact, this mode of thought that started the social sciences along a blind path in the first place, and that despite thousands of elaborations since, the fundamental assumptions expressed by Durkheim still form the underpinning of social science theory (e.g., Rex, 1961; Evans-Pritchard, 1954; Harris, 1979).[1]

And, although I have chosen my own fields of sociology and social anthropology for particular scrutiny, I also note that, even in such fields as economics, major models are still built around the assumption that executives within a corporation are working for the good of the corporation. I leave to the reader to judge to what extent such models are consistent with reality.

RESISTANCE TO SOCIOBIOLOGY BY THE SOCIAL SCIENCES

Although the burgeoning field of vertebrate behavioral ecology now has a very large agenda, its original impetus came through the recognition that fitness consists not just of the reproduction of one individual's genes ("individual fitness"), but also of all the genes that an individual shares by common descent with relatives—that is, "inclusive fitness" or "kin selection." This single insight, with its many ramifications, brilliantly explained much of the enormous variety of interactions, dispersal, and formation of social groups of organisms of every kind. Coincidental with these developments in theory, long-term field studies of various animals, especially birds and primates, had reached the point in the mid-1970's where individual animals had been observed through most or all of their life cycle. There was, therefore, ready and abundant proof that social interactions were structured by kinship (e.g., Goodall, 1986; Smuts, 1985; Smuts et al., 1986).

The application of this theory and evidence to small-scale human societies seemed to me immediate and obvious: as every graduate student in social anthropology learns, "the natives are obsessed with kinship." Among various groups where I have worked, one cannot even have a conversation until one has been placed firmly within a kinship constellation; there is no alternative except "stranger" (and therefore, potentially, "enemy"). When I proudly announced to my colleagues in social anthropology that, in their vast libraries on kinship and social organization, they

---

[1] Consider the following passage from the essay "Social Anthropology," by E. E. Evans-Pritchard, Professor of Social Anthropology at Oxford, and one of the most influential figures of the modern period:

"Durkheim's importance in the history of the conceptual development of social anthropology in this country might have been no greater than it has been in America had it not been for the influence of his writings on Professor A. R. Radcliffe-Brown and the late Professor B. Malinowski, the two men who have shaped social anthropology into what it is in England today...

"Radcliffe-Brown has...clearly and consistently stated the functional, or organismic, theory of society...'the concept of function applied to human societies is based on an analogy between social life and organic life.' Following Durkheim, he defines the function of a social institution as the correspondence between the social institution and the necessary conditions of existence of the social organism...so conceived of, a social system has a functional unity. It is not an aggregate but an organism or integrated whole (1954: 53–4)."

had by far the largest body of data to contribute to this emerging new paradigm, they were appalled; they had come to define human kinship as purely cultural, symbolic and arbitrary, with little or no relationship to the "biological facts" of kinship. They were quite wrong, of course. But the attack on sociobiology, as it came to be popularly called after the publication of Wilson's influential volume (1975), was immediate, immoderate and immense (e.g., Sahlins, 1976). Paralleling the burgeoning libraries on sociobiology, there is a modest growth industry in "critiques of sociobiology." The reasons for these attacks are largely based on historical antagonisms and often have little bearing on the actual data and theory of sociobiology. Here are some of them:

First, there is the familiar and well-known fear of "reductionism" in the social sciences. This is hardly an irrational fear; even the most cursory look at the history of the social sciences will show how frequently they have been buffeted by arrogant biological argument. In the majority of cases (race differences, eugenics, immigration laws, I.Q. testing, etc.), the effect has been to brutalize social reality and minimize environmental influences, often for the most transparent and self-serving reasons (e.g., Kevles, 1985). There has, thus, developed an antagonism toward reductionism of any kind so strong that most of you at this meeting would have difficulty comprehending it. As Lionel Tiger has remarked, "If, in the physical sciences, one is able to successfully reduce complex phenomena to a simple rule or model, one is awarded the Nobel Prize; the reward for a similar attempt in the social sciences is to be pilloried in the New York Review of Books." Although they would not put it so crudely, my colleagues in social and cultural anthropology seem to be comfortable with the belief that biology and natural selection successfully delivered *Homo sapiens* into the upper Paleolithic, and then abandoned our species to the pure ministrations of culture. (There has recently been a spate of promising attempts to model genetic/cultural co-evolution, but these have come from biologists and biological anthropologists, not social and cultural ones [Lumsden and Wilson, 1981; Feldman and Cavalli-Sforza, 1979; Durham, in press].)

From my point of view, such attitudes threaten to leave the social sciences on a very small and sandy island in a rapidly rising river: theory that has now been shown to apply to plants, single cells, vertebrates, and invertebrates (Trivers, 1985) would have to exempt humans; theory that is acknowledged to apply during the first 8 million years of our evolution would be considered inapplicable to the last 30 or so thousand years (Tooby and DeVore, 1987). In fact, natural selection— properly understood as differential reproductive success—was greatly *accelerated* in the period following our "hunter-gatherer stage" of evolution. It was only after societies had begun to develop significant social stratification, caste and class, that polygyny become commonplace, and selection could then be accelerated through the major differences in reproductive success between high and low status males.

THE ISSUE OF AGGRESSION

Another reason for social scientists to reject sociobiology is that many of the most accessible writings on behavioral biology and evolution in recent decades have themselves been seriously muddled. For example, Robert Ardrey (1961) popularized the thesis that, because of the purported hunting-killing-cannibalistic way of life of our hominid ancestors, humans had inherited ineradicable instincts for violence and warfare—a chilling scenario brought vividly to life by Stanley Kubrick in "2001: A Space Odyssey." Ardrey's *African Genesis* was singled out by *Time* as "one of the ten most influential books of its decade." Konrad Lorenz, a Nobel Laureate, advanced a quite different and more careful argument in his popular *On Aggression* (1966). Lorenz observed that wherever he looked in the animal kingdom, he found aggressive competition. Aggression, he reasoned, must be a *sine qua non* of life; the structures and behaviors of aggression are "necessary if only the fittest are to survive, mate successfully, and carry on the species." Animal aggression, however, was most often expressed by bluff and ritualized combat; mortal wounds were rare. Since aggression is also inevitable in the human species, he felt that our best hope lay in finding more constructive ways to channel and release our aggressive impulse.

Lorenz's argument rested on several faulty assumptions. Biologists no longer believe that individuals are behaving "for the good of the species." Furthermore, even if this were the case, Lorenz was using a narrow and discredited definition of "fitness"—one that equates fitness with strength and superior fighting ability. While biologists believe that the evolution of a species by natural selection depends upon competition within that species, they do not believe that success in such competition is measured by either strength or longevity; the ultimate test of fitness is reproductive success. More precisely, when we assess the fitness of an individual (or a gene or a behavior), we now look beyond the individual animal to also consider the effects on the fitness of the individual's kin. Kin selection, or inclusive fitness, considers both the consequences of any behavior upon one's own reproduction, and also the consequences for the reproductive success of one's kin—that is, individuals with whom one shares genes by common descent.

From this point of view, one may ask whether the inclusive fitness of an individual will or will not be best advanced by an act of aggression; but, in any case, one should not assume that aggression is contributing to the fitness or success of a whole species. On the contrary, consider the enormous energy investment an individual must make in order to be aggressive: energy must be diverted to building muscles, claws, tusks or horns, leading to a high ontogenetic cost and resulting in delayed maturation—and all this *before* expending energy in the act of aggression itself. If we could somehow redesign the evolutionary process, we would probably conclude that a species would be far better off if it could simply dispense with these costs and invest the energy in more beneficial pursuits—e.g., in better quality care for the immature members of the group. In the real world, the "aggressive complex" of morphology and aggressive behavior, which promotes successful reproduction for oneself and one's relatives, will probably *lower* the chances for survival of the group, population or species (Konner, 1982).

We also now know that ritualistic combats are only part of the aggression story. Numerous decade-long studies of animal behavior show that animal murder and infanticide are not rare events. (Ironically, the human species may not be the "killer-apes" Ardrey supposed, but, in fact, among the more pacific species.) We now realize that ritualized aggressive encounters are better explained by models such as those Maynard Smith has advanced as "evolutionarily stable strategies" (1982). For example, if two opponents can determine by some non-lethal means which one would win an all-out fight, it would be advantageous to both the *winner* and the *loser* to determine this outcome in advance, by bluff and tests of strength, without bloodshed (Popp and DeVore, 1979).

Many of us look back rather wistfully on the notion that, for altruistic reasons, animal were deliberately handicapping themselves and substituting ritual for real combat, but the facts argue otherwise. I do not intend, by this long example, to discredit Lorenz's other major achievements; he is one of the giants of modern biology. On the contrary, I mean to illustrate how even the very best of Darwin's descendants were severely handicapped by the state of theory in evolutionary biology only 20 years ago.

THE GROUP SELECTION FALLACY

The compelling logic of group or species-advantage theory dies very hard; even some biologists continued to defend it into the early 1960's. The theory found its most articulate spokesman in Wynne-Edwards, who held that, by various behavioral mechanisms, species practiced "prudential restraint" on reproduction and that "it must be highly advantageous to survival, and thus strongly favoured by selection, for animal species (1) to control their own population densities and (2) to keep them as near as possible to the optimum level for each habitat" (1962:9)—the "optimum level" being below that at which food resources would be depleted and the population crash. He continues: "Where the two [group selection and individual selection] conflict, as they do when the short-term advantage of the individual undermines the future safety of the race, group-selection is bound to win, because the race will suffer and decline, and be supplanted by another in which antisocial advancement of the individual is more rigidly inhibited" (1962:20).

Attractive as this "prudential" line of reasoning may be, Wynne-Edwards' large volume was soon savaged by theoretical biologists and his examples refuted in detail; his argument has had no credibility in evolutionary biology since the mid-60's. But, such is the estrangement between biology and social science, that, in the same years in which Wynne-Edwards' arguments were being discarded in biology, Roy Rappaport turned to this very work for the theoretical underpinning of his highly praised book, *Pigs for the Ancestors* (1967). This study was correctly viewed as the most sophisticated book on human cultural ecology to yet appear. Rappaport had used admirably quantitative data to support his analysis of food resources, social organization, warfare, and the ritual cycle—but the theory to which he referred his analysis had already been discredited. Nor has the pernicious influence of group-selection thinking abated; see, for example, any of the many works of Marvin Harris (e.g., 1979).

I would consider myself high on the list of those who would welcome assurances from theoretical biology that what J. B. S. Haldane called "Pangloss's theorem" ("all is for the best...") is correct. This is not the place to detail the evidence against group selection arguments, but if one would argue that individuals have been selected to behave in a "group" or "species-altruistic" manner, then one must wonder why the paleontological record shows that 99% of all species no longer exist; natural selection has condemned most of them to extinction, and a large portion of the remainder are so changed that we cannot even determine the ancestral form. Clearly, selection for behavior that would benefit the group or species has consistently lost out to selection for behavior that benefits the more selfish genetic interests of the individual.

GENETIC DETERMINISM

Finally, but by no means least important, it must be admitted that the aggressive, aggrandizing stance of many of the early converts to sociobiology were hardly designed to put social scientists at ease. This was due in part to the theoretician's penchant for proposing a change in a single gene as a way of modeling a behavioral change that all of us would acknowledge was, in fact, a far more complex reality at both the genetic and behavioral levels. But the pattern had been set by W. D. Hamilton's original formulations (1964), which had ascribed behavioral evolution to genetic mutation, and subsequent models have also begun with such assumptions as: "suppose an altruistic mutant for kin-directed altruism appears in a population of selfish individuals." Such a model then attempts to specify the conditions under which the gene for such behavior will spread at the expense of its alternate alleles. Detailed fieldwork on a wide range of species has largely confirmed the expectations of such models. But when Dawkins carried such reasoning to its logical conclusion in his lucid, witty (and best-selling) *The Selfish Gene* (1976), his metaphor for genetic replication, whatever its heuristic value, was too graphic for any but the true believer: "Now they [replicating genes] swarm in huge colonies, safe inside gigantic lumbering robots, sealed off from the outside world, communicating with it by tortuous indirect routes, manipulating it by remote control. They are in you and in me; they created us, body and mind; and their preservation is the ultimate rationale for our existence."

Now that the paradigm shift is complete, and evolutionary biology and vertebrate ecology have been permanently altered, the impetuousness of the early revolutionaries has given way to the more mundane pursuits of normal science. Indeed, one indication of the health and vigor of sociobiology is that it has tended to be self-correcting. It was the trenchant critique of a fellow sociobiologist that exposed the naive and simplistic approach to human behavior by some early practitioners of "human sociobiology" (Dickemann, 1979). Another sociobiologist has argued that "the focus on genetic mutations, which so advanced the field for so many years is now constraining it...behavioural mutants can arise and spread through a population even in the absence of a causal genetic change" (Wrangham, 1980). Sarah Hrdy, at once a feminist and a sociobiologist, has been instrumental in exposing the androcentric biases in studies of primate behavior and reconstructions of hominid evolution (1981).

PROSPECTS FOR A SYNTHESIS

The structural-functional paradigm in the social sciences has been moribund for decades. The harmonious, static model of society it offered has proven to be completely incapable of dealing with such dynamics of social organization as social change, intra-group competition, sexual politics, and parent-child conflict. The invention of weak concepts such as "social dysfunction" are symptomatic of the attempt to shore up a dying paradigm. One result has been that even Marxist analysis has at last been given a fair hearing. But, however vigorously the partisans of these various social theories may contend among themselves, I believe that they are all engaging in superficial argument, and are failing to address the real crisis in fundamental social theory (e.g., Harris, 1979).

I do not for a moment expect that most social scientists will now turn to evolutionary biology for enlightenment. The painful and ironic history of attempts by social scientists to borrow metaphor and theory from biology will intimidate all but the most daring. I have singled out group selection theory as the primary villain of this piece because it is so easy to show its pernicious influence, in both biology and the social sciences, but the revolution in behavioral biology goes far beyond the mere expunging of this concept; a whole spectrum of theory and methodology in evolutionary biology has now been sharpened and clarified (see Tooby, this volume, and references in Trivers, 1985).

The rapid advances in vertebrate behavioral ecology have grown out of the constant dialogue between theory and testing; theory is quickly refined and used to test new hypotheses in the field and laboratory. The theory and methods that have been developed are excellent instruments for analyzing the interactions of organisms in face-to-face encounters, and it is in the description and interpretation of such human interactions that they will have the most impact on the human behavioral sciences. (I do not foresee any direct use of sociobiology in many areas of human inquiry—for example, the explication of ethnohistory, or the details of religious ritual.)

To achieve even a modest synthesis in the human sciences, however, most social scientists will have to radically change their methods of gathering and analyzing data. Much of the success in behavioral biology has come through painstakingly detailed descriptions of the interactions of individual organisms over most or all of the life cycle. My initial enthusiasm for the "large body of data" available from a century of anthropological fieldwork was quite misplaced. With notable exceptions, most anthropological monographs have reduced the real data on human social life to idealized or averaged patterns of behavior and belief. For example, alternative kinship terms, which are used in the real world to negotiate social relationships, have most often been treated as sloppy deviations around an idealized terminology system. That is, the anthropologist has considered it necessary to reduce the "confusion of data" to a single, coherent terminological system. But this approach simply parallels the fundamental error that plagued biology for so long—the tendency to see the world as a typologist, rather than as a populationist. Ernst Mayr states the constrast very well: "The ultimate conclusions of the population thinker and of the typologist are precisely the opposite. For the typologist, the type (*eidos*) is real and the variation an illusion, while for the populationist the type (average) is an abstraction and only the variation is real. No two ways of looking at nature could be more different" (1976:28).

CONCLUSION

Much of this essay has been negative because I have sought to explain in a few pages why the majority of social scientists continue to vigorously resist any rapprochement with behavioral biology. But, in fact, many younger behavioral scientists are already working within that framework (e.g., Chagnon and Irons, 1979). Field teams are beginning to collect the kinds of data that will allow human behavior to be tested against the same hypotheses that have informed the study of other vertebrates. Many books and new journals are now devoted largely or entirely to "human sociobiology" (e.g., Alexander, 1979, 1987; Daly and Wilson, 1987; Konner, 1982). My own interest, and the subject of this essay, has been the development of theory and method as they are transforming field studies of humans and other vertebrates. I am confident that new insights will emerge when human behavior can be treated in the same framework that we apply to other animals (see Wrangham, this volume). But this same approach is also capable of providing a coherent and deductive framework for other human sciences, such as psychology, as indicated in the following essay by John Tooby.

We have only begun to explore the deductive power and implications of this emerging synthesis.

ACKNOWLEDGMENT

I thank Robert Trivers for first explaining to me the revolution in evolutionary biology, John Tooby for many conversations relating to the topics treated here, and Murray Gell-Mann for the dubious distinction of convincing me that my informal remarks at the conference deserved wider circulation. John Tooby and I were not able to collaborate on this paper, as planned, and his contribution follows. Nancy DeVore and Nancy Black were, as always, extraordinarily patient and helpful in the manuscript preparation. This work was supported in part by NSF Grant No. BNS-83-06620 and the L. S. B. Leakey Foundation

REFERENCES

Alexander, R. D. (1979), *Darwinism and Human Affairs*, London: Pitman Publishing Limited.

Alexander, R. D. (1987), *The Biology of Moral Systems*, New York: Aldine de Gruyter.

Ardrey, Robert (1961), *African Genesis*, New York: Atheneum.

Chagnon, Napoleon A. and William A. Irons, editors (1979), *Evolutionary Biology and Human Social Behavior: An Anthropological Perspective*, North Scituate, MA: Duxbury Press.

Daly, Martin and Margo Wilson (1987), *Homocide*, New York: Aldine de Gruyter.

Dawkins, R. (1976), *The Selfish Gene*, New York: Oxford University Press.

Dawkins, R. (1982), *The Extended Phenotype: The Gene as the Unit of Selection*, Oxford: Oxford University Press.

Dickemann, Mildred (1979), "Comment on van de Berghe and Barash's Sociobiology," *American Anthropologist* **81**, No. 2 (June, 1979): 351–57.

Durham, William (in press), *Coevolution: Genes, Culture and Human Diversity*, Stanford: Stanford University Press.

Durkheim, E. (1895), *Les Regles de la Methode Sociologique*, Paris. (Translation 1938, *The Rules of Sociological Method*, Sarah A. Solway and John H. Mueller, Glencoe, Illinois: Free Press.)

Evans-Pritchard, E. E. (1954), *Social Anthropology*, Glencoe, Illinois: Free Press.

Feldman, M. W. and L. L. Cavalli-Sforza (1979), "Aspects of Variance and Covariance Analysis with Cultural Inheritance," *Theoretical Population Biology*, **15** (3): 276–307.

Goodall, Jane (1986), *The Chimpanzees of Gombe: Patterns of Behavior*, Cambridge, MA: Harvard University Press.

Hamilton, W. D. (1964), "The Genetical Evolution of Social Behavior," I and II, *Journal of Theoretical Biology*, **7** (1): 1–52.

Harris, Marvin (1979), *Cultural Materialism,* New York: Random House.

Hrdy, S. B. (1981), *The Woman that Never Evolved,* Cambridge, MA: Harvard University Press.

Kevles, D. J. (1985), *In the Name of Eugenics: Genetics and the Use of Human Heredity,* New York: Alfred A. Knopf.

Konner, Melvin (1982), *The Tangled Wing: Biological Constraints on the Human Spirit,* New York: Harper and Row.

Krebs, J. R. and N. B. Davies (1984), *Behavioral Ecology: An Evolutionary Approach,* 2nd Edition, Sunderland, MA: Sinauer Assoc. Inc.

Lee, Richard B. and Irven DeVore, editors (1976), *Kalahari Hunter-Gatherers: Studies of the !Kung San and Their Neighbors,* Cambridge, MA: Harvard University Press.

Lorenz, Konrad (1966), *On Aggression,* London: Methuen.

Lumsden, Charles and E. O. Wilson (1981), *Genes, Mind and Culture: The Co-evolutionary Process,* Cambridge, MA: Harvard University Press.

Maynard Smith, John (1964), "Group Selection and Kin Selection," *Nature,* London **201** (4924): 1145–1147.

Mayr, Ernst (1976), *Evolution and the Diversity of Life,* Cambridge, MA: Harvard University Press.

Popp, Joseph L. and Irven DeVore (1979), "Aggressive Competition and Social Dominance Theory," *The Great Apes,* Eds. D. A. Hamburg and E. R. McCown, Menlo Park, CA: The Benjamin/Cummings Publishing Co., 317–340.

Rappaport, Roy (1967), *Pigs for the Ancestors,* New Haven: Yale University Press.

Rex, John (1961), *Key Problems of Sociological Theory,* New York: Humanities Press.

Rubenstein, d. I. and R. W. Wrangham, eds. (1986), *Ecological Aspects of Social Evolution,* Princeton University Press.

Sahlins, Marshall (1976), *The Use and Abuse of Biology: An Anthropological Critique of Sociobiology,* Ann Arbor: University of Michigan Press.

Shostak, Majorie (1981), *Nisa: The Life and Words of a !Kung Woman,* Cambridge, MA: Harvard University Press.

Smuts, Barbara B. (1985), *Sex and Friendship in Baboons,* New York: Aldine de Gruyter.

Smuts et al., eds. (1986), *Primate Societies,* University of Chicago Press.

Tooby, John and Irven DeVore (1987), "The Reconstruction of Hominid Behavioral Evolution through Strategic Modeling," *The Evolution of Human Behavior: Primate Models,* Ed. W. G. Kinzey, Albany, NY: State University of New York Press, 183–237.

Trivers, Robert (1985), *Social Evolution,* Menlo Park, CA: The Benjamin/Cummings Publishing Company.

Williams, G.C. (1966), *Adaptation and Natural Selection: A Critique of Some Current Evolutionary Thought,* Princeton, New Jersey: Princeton University Press.

Wilson, E. O. (1975), *Sociobiology: The New Synthesis,* Cambridge, MA: Belknap Press of Harvard University Press.

White, Leslie (1949), *The Science of Culture,* New York: Grove Press.

Wrangham, Richard (1980), "Sociobiology: Modification with Dissent," *Biological Journal of the Linnaean Society* **13**, 171–177.

Wynne-Edwards, V. C. (1962), *Animal Dispersion in Relation to Social Behavior,* Edinburgh: Oliver and Boyd.

JOHN TOOBY
Department of Anthropology, Harvard University

The Emergence of Evolutionary Psychology

Humans, like all other organisms, were created through the process of evolution. Consequently, all innate human characteristics are the products of the evolutionary process. Although the implications of this were quickly grasped in investigating human physiology, until recently there has been a marked resistance to applying this knowledge to human behavior. But evolution and the innate algorithms that regulate human behavior are related as cause and consequence: lawful relations are being discovered between the evolutionary process and the innate psychology it has shaped. These lawful relations constitute the basis for a new discipline, evolutionary psychology, which involves the exploration of the naturally selected "design" features of the mechanisms that control behavior. This synthesis between evolution and psychology has been slow in coming (see DeVore, this volume). The delay can be partly accounted for by two formidable barriers to the integration of these two fields: the initial imprecision of evolutionary theory, and the continuing imprecision in the social sciences, including psychology.

The revolution in evolutionary theory began two decades ago and, gathering force, has subsequently come to dominate behavioral inquiry. Vague and intuitive notions of adaptation, frequently involving (either tacitly or explicitly) group selection, were replaced by increasingly refined and precise characterizations of the evolutionary process (Williams, 1966; Maynard Smith, 1964; Hamilton, 1964). The application of these more precise models of selection at the level of the gene opened the door for meaningful explorations of a series of crucial behavioral problems, such as altruism towards kin, aggression, mate choice, parental care, reciprocation, foraging, and their cumulative consequences on social structure. These theoretical advances had their most dramatic impact on field biology, quickly reorganizing research priorities, and integrating the diverse studies of animal (and plant) behavior into a larger system of evolutionarily-based behavioral ecology (or sociobiology).

The heart of the recent revolution in evolutionary theory lies in the greater precision with which the concept *adaptation* is now used: the primary evolutionary explanation for a trait is that it was selected for; this means that it had or has the consequence of increasing the frequency of the genes that code for it in the population; if there is such a correlation between a trait and its consequences, the trait can then be termed an *adaptation*; the means by which a trait increases the frequency of its genetic basis is called its *function*. There is no other legitimate meaning to adaptation or function in the evolutionary lexicon. Thus, the genes present in any given generation are disproportionately those which have had, in preceding environments, effective "strategies" for their own propagation. The traits individuals express are present because the genes which govern their development were incorporated in the genome because they have successful strategies of self-propagation. In other words, genes work through the individual they occur in, and the individual's morphology and behavior embody the strategies of the genes it contains.

The conceptual vagueness of the theory of natural selection, as it existed before these advances, meant that psychologists found little in it that they could meaningfully apply to produce coherent behavioral theories. However, instead of the earlier vague and impressionistic accounts of adaptation, modern behavioral ecology supplies a cogent set of specific predictions that are straightforwardly derived from a validated deductive framework. The mathematical and conceptual maturation of evolutionary theory has therefore removed one of the principal barriers to the creation of a coherent evolutionary psychology.

The second conceptual impediment has been the vagueness of psychology itself, both in its formulation of theories, and in its description of psychological phenomena. The field has floundered in a sea of incompatible and inchoate theories and interpretive frameworks since its inception. Despite the crippling limitations of the behaviorist paradigm, it is easy to sympathize with the driving motivation behind it: impatience and frustration with the incoherence and uninformativeness of unspecified and impressionistic assertions, theories, and descriptions. The rapid development of modern computer science, however, has begun to transform the field of psychology, especially in the last fifteen years. The capacity to specify intricate information-based dynamical procedures both legitimized and made feasible the construction of rigorously specified models of how humans process information. The creation of cognitive psychology has been one consequence.

The methodological advances and insights of cognitive psychology have cleared away the last conceptual impediment to the development of an integrated evolutionary psychology by providing an analytically precise language in which to describe behavior-regulating algorithms. In fact, the "algorithmic" language of cognitive psychology and behavioral ecology dovetail together: *strategies* defined by ecological theory are the analytical characterizations of the selective forces that have shaped the proximate mechanisms that collectively comprise the psyche. The concepts (and technology) of computer science allow the formulation of dynamical decision structures and procedures that can tightly model the psychological algorithms which actually control behavior, guiding it onto adaptive paths. Starting from the realization that all of the psychological mechanisms are there *solely* because they evolved to promote the inclusive fitness of the individual, researchers can, for the first time, correctly understand the function of human psychological characteristics. Knowing the function of psychological mechanisms provides a powerful heuristic for defining them, investigating them, and evaluating hypotheses about their architecture.

As a result, the potential for advances in evolutionary psychology is beginning to be realized. The only remaining limitations are institutional: the psychological research traditions which antedate these advances in evolutionary theory remain insulated from and largely ignorant of their important uses and implications. There remains, of course, considerable vested interest in a corpus of research whose interpretive basis rests on obsolete assumptions.

ARE HUMANS IMMUNE TO BEHAVIORAL EVOLUTION?

This institutional resistance is manifested by the prevalent belief that, while evolution shaped other species' psyches, it is irrelevant to human behavior, because of the existence of culture, intelligence, and learning. Thus, the argument runs, in the transition from simpler primate behavioral mechanisms, to the more elaborated and powerful ones we know to be present in modern humans, a crucial boundary was crossed. Many regard this, almost mystically, as a watershed transition which places human phenomena in another category entirely, beyond the capacity of evolutionary and ethological methods to study, model, or understand. They take the uniqueness of humanity (which is undoubted) to mean its incomprehensibility in evolutionary terms (e.g., Sahlins, 1976).

However, the immense increase in complexity of human (and protohuman) behavior is tractable to evolutionary psychology. Essential to evolutionary modeling is the distinction between proximate means and evolutionary ends. What proximate mechanisms are selected ("designed") to accomplish is the promotion of inclusive fitness. This end is fixed and is intrinsic to the evolutionary process. The mechanisms by which fitness is promoted may change over evolutionary time. However, the elaboration of mechanisms from the simple into the complex changes only the proximate means, not the evolutionary ends. In fact, such changes will occur only when they increase inclusive fitness, that is, only when they better promote the same evolutionary ends. Humans are characterized by a remarkable expansion in intelligence, consciousness (however defined), complex learning and culture transmission mechanisms, all interpenetrated by a sophisticated coevolved motivational system. But evolutionary psychology is uniquely suited to the analysis of these mechanisms, precisely because it analyzes mechanisms in terms of evolutionary ends, which do not change. As intelligence, learning, consciousness, and motivational systems progressively become more sophisticated, they still serve the same strategic ends according to the same evolutionary principles (Tooby & DeVore, 1987).

Those who continue to assert that humans became immune to the evolutionary process, and are not significantly shaped by evolutionary principles, must somehow reason their way past the following fatal objection to both sophisticated and simple versions of their position. The innate characteristics whose genetic basis has become incorporated into the human genome were incorporated because they increased inclusive fitness, and therefore they are adaptively patterned. To assert anything else is to maintain that somehow a large number of less fit innate characteristics (those which did not correlate with fitness) displaced those that were more fit. In other words, they are faced with explaining how evolutionary processes systematically produced maladaptive traits. Usually, this kind of thinking is based on the notion that culture replaces evolution, and has insulated human behavior from selective forces. However, the existence of culture can only mean that natural selection produced and continues to shape the innate learning mechanisms which create, transmit, and assimilate cultural phenomena. These innate learning mechanisms, as well as their associated innate motivational, emotional, and attentional systems, control what humans choose to learn, what sorts of behavior they find reinforcing, and what goals they pursue, rather than the precise means by which they pursue them. Humans are unique in means, not in ends. The residual sense in the cultural insulation argument is the sound but simple one of phylogenetic lag: modern humans have emerged so rapidly from Pleistocene conditions that their mechanisms are still following the programming of what would have been adaptive under Pleistocene conditions.

In fact, sophisticated hominid mechanisms, instead of being divergent from evolutionary principles, may more purely incarnate adaptive strategies. Hominids' more intelligent, flexible, and conscious systems are less limited by mechanistic and informational constraints, and can more sensitively track special environmental, historical, and situational factors and make appropriate adaptive modifications. Evolutionary processes select for any behavioral mechanism or procedure, no matter how flexible or how automatic, that correlates with fitness.

The set of behaviors which lead to survival and genetic propagation are an extremely narrow subset of all possible behaviors. To be endowed with broad behavioral plasticity is an evolutionary death sentence unless this plasticity is tightly bound to a "guidance system" which insures that out of all possible behaviors, it is those that promote inclusive fitness which are generated. Selection for plasticity must have been linked to the development of such a sophisticated guidance system in humans, or it could never have occurred. In fact, the primary task of human evolutionary psychology is the elucidation of this constellation of guiding algorithms. The existence of this guidance system prevents the "escape" of human behavior from analysis by evolutionary principle. Evolutionary psychology is not thwarted by hominid singularity. Evolutionary analysis shows hominid uniqueness to be rule-governed rather than imponderable. While it may prove that many hominid adaptive elements are combined in novel ways, this does not mean they are put together in random or unguessable ways.

SOME EARLY SUCCESSES IN EVOLUTIONARY PSYCHOLOGY

Despite the fact that cognitive psychology has developed, by in large, uninfluenced by evolutionary biology, the realities of the human mind are forcing cognitive psychologists towards many of the same conclusions implicit in the evolutionary approach. Researchers in artificial intelligence have been chastened in their attempts to apply cognitive theory to produce actual (computational) performance. Simple associationistic theories of learning proved completely inadequate. They discovered that in order to get a system to do anything interesting (such as "see," learn syntax, analyze semantic content, manipulate objects in a three-dimensional world, etc.), they had to provide the program with massive amounts of specific information about the domain the program was supposed to learn about or manipulate; in other words, they had to give the computer a great deal of "innate knowledge." This phenomenon is so pervasive and so well-recognized that it has a name: the frame problem (Boden, 1977). Moreover, the program had to contain highly structured procedures specialized to look for exactly those types of relationships which characterized the problem domain. Such procedures correspond to innate algorithms in the human psyche. It was possible to be an extreme environmentalist only as long as the researcher was not forced to get too specific about how performance was actually achieved. In artificial intelligence, this was no longer possible.

These realizations were foreshadowed by developments in psycholinguistics. Because syntax constituted a formally analyzable system, Chomsky was able to show that humans must have a powerful innate language-acquisition device in order to learn it. In Chomsky's phrase, the stimuli (the utterances of adults) were too impoverished to provide sufficient information for a child to learn the correct grammar through induction (Chomsky, 1975; Wanner & Gleitman, 1982). Humans had to have innate expectations or algorithms constraining the possible set of grammatical relations. This led Chomsky to beliefs similar to those implicit in evolutionary psychology: that the mind is composed of "mental organs" just as specialized in function as our physiological organs are.

By recognizing that the mind includes domain-specific algorithms or modules which are "designed" for or adapted to specific purposes, rapid progress has been made on a number of problems. For example, Marr (1982) uncovered the outlines of how the mind constructs three-dimensional objects from a two-dimensional retinal array. Roger Shepard, reasoning soundly from evolutionary principle, has demonstrated that the algorithms that govern our internal representations of the motions of rigid objects instantiate the same principles of kinematic geometry that describe the motion of real objects in the external world. Experimental evidence from perception, imagery, apparent motion, and many other psychological phenomena support his analysis (Shepard, 1984). As he points out, "through biological evolution, the most pervasive and enduring constraints governing the external world and our coupling to it are the ones that must have become most deeply incorporated into our innate perceptual machinery" (Shepard, 1981). Motivated by similar considerations, Carey and Diamond (1980) provide persuasive evidence from a wide array of psychological and neurological sources that humans have innate face-encoding mechanisms. Daly and Wilson, in a series of important studies, have found strong evidence indicating evolutionary patterning in such diverse phenomena as homicidal behavior, differential parental care, and sexual jealousy (Daly & Wilson, 1980; 1981; 1982; Daly, Wilson & Weghorst, 1982).

The extensive literature on human reasoning provides an opportunity for the demonstration of the usefulness of the evolutionary approach. Research on logical reasoning showed that humans frequently reasoned illogically, when the standard for valid reasoning was adherence to formal logic (Wason and Johnson-Laird, 1972). The conclusions people arrive at vary widely depending on the specific content they are asked to reason about. Research on so-called content effects in logical reasoning has been bogged down in a quagmire of conflicting results and interpretations, and none of the prevailing hypotheses have demonstrated any predictive power.

Cosmides (1985) has productively reorganized this confused literature through the application of the evolutionary approach. The content effects become very orderly when they are scrutinized for the presence of evolutionarily significant content themes. Psychological mechanisms evolved to handle important and recurrent adaptive problems (such as face recognition, mentioned above), and one crucial adaptive problem for humans is social exchange. Trivers (1971) and Axelrod & Hamilton (1981) demonstrated that cooperation can evolve only if individuals identify and bestow benefits on those likely to reciprocate and avoid such deferred exchange relationships with those who "cheat" through inadequate reciprocation. Because such cooperative labor and food-sharing exchanges have typified human hunter-gatherer bands throughout their evolutionary history, humans have depended on the evolution of a cognitive/motivational mechanism that detects potential cheaters in situations involving social exchange. Cosmides (1985) showed that an adaptive logic designed to look for cheaters in situations of social exchange predicts performance on logical reasoning tasks which involve such social content. Her elegant series of experiments have provided solid support for the hypothesis that humans have an innate special-purpose algorithm which structures how they reason about social exchange, with properties that differ markedly from formal logic. Not only do humans have an innate language-acquisition device, but they appear to have a collection of innate inferential networks which structure their reasoning about the social world.

Indeed, the evolutionary approach contains the potential for clarifying the murky area of emotion, and its relation to cognition (Tooby & Cosmides, in press). If the mind is viewed as an integrated architecture of different special-purpose mechanisms, "designed" to solve various adaptive problems, a functional description of emotion immediately suggests itself. Each mechanism can operate in a number of alternative ways, interacting with other mechanisms. Thus, the system architecture has been shaped by natural selection to structure interactions among different mechanisms so that they function particularly harmoniously when confronting commonly recurring (across generations) adaptive situations. Fighting, falling in love, escaping predators, confronting sexual infidelity, and so on, have each recurred innumerable times in evolutionary history, and each requires that a certain subset of the psyche's behavior-regulating algorithms function together in a particular way to guide behavior adaptively through that type of situation. This structured functioning together of mechanisms is a mode of operation for the psyche, and can be meaningfully interpreted as an emotional state. The characteristic feeling that accompanies each such mode is the signal which activates the specific constellation of mechanisms appropriate to solving that type of adaptive problem.

To make this concrete, let us briefly describe in these terms what might happen to a hypothetical human hunter-gatherer when a distant lion becomes visible. The recognition of this predator triggers the internal "broadcast" of the feeling of fear; this feeling acts as a signal to all of the diverse mechanisms in the psychological architecture. Upon detecting this signal, they each switch into the "fear mode of operation": that is, the mode of operation most appropriate to dealing with danger presented by a predator. The mechanism maintaining the hunger motivation switches off and cognitive activity involved in reasoning about the discovery of food is stopped, neither being appropriate. A different set of motivational priorities are created. Mechanisms regulating physiological processes issue new "instructions" making the person physiologically ready for the new sorts of behaviors which are now more adaptive: fighting or, more likely, flight. Cognitive activity switches to representations of the local terrain, estimates of probable actions by the lion, sources of help and protection from the lion, and so on. The primary motivation becomes the pursuit of safety. The modes of operation of the perceptual mechanisms alter radically: hearing becomes far more acute; danger-relevant stimuli become boosted, while danger-irrelevant stimuli are supressed. The inferential networks underlying the perceptual system interpret ambiguous stimuli (i.e., shadows, masking noise) in a threatening way, creating a higher proportion of false positives. Attention-directing mechanisms become fixed on the danger and on potential retreats.

In this view, emotion and cognition are not parallel processes: rather, emotional states are specific modes of operation of the entire psychological architecture. Each emotional state manifests design features "designed" to solve particular families of adaptive problems, whereby the psychological mechanisms assume a unique configuration. Using this approach, each emotional state can be mapped in terms of its characteristic configuration, and of the particular mode each identifiable mechanism adopts (motivational priorities, inferential algorithms, perceptual mechanisms, physiological mechanisms, attentional direction, emotion signal and intensity, prompted cognitive contents, etc.).

Evolutionary psychology employs functional thinking, that is, the modern rigorous understanding of adaptive strategies, to discover, sort out, and map the proximate mechanisms that incarnate these strategies. In so doing, it appears to offer the best hope for providing a coherent and unified deductive framework for psychology. Sciences make rapid progress when they discover the deductive framework that is appropriate to their phenomena of study. Fortunately, there exists in biology a set of principles with the requisite deductive power: evolutionary theory. We know that humans evolved, and that the mechanisms that comprise our psyches evolved to promote fitness. Our innate psychological algorithms are rendered comprehensible by relating them to a rigorously characterized evolutionary process. These realizations can organize research efforts in psychology into valid and productive investigations, because evolutionary analysis provides the level of invariance that reveals behavioral variation to be part of an underlying system of order (Cosmides, 1985; Tooby and DeVore, 1987; Cosmides and Tooby, 1987; Tooby and Cosmides, in press).

ACKNOWLEDGMENT

This article was to have been part of a joint effort by Irven DeVore (this volume) and myself. Unfortunately, circumstances prevented us from consolidating our two halves. Nevertheless, I gratefully acknowledge his valuable assistance, and equally thank L. Cosmides for her deep and extensive participation.

REFERENCES

Axelrod, R. & Hamilton, W. D. (1981). "The Evolution of Cooperation." *Science* **211**, 1390–96.

Boden, M. (1977). *Artificial Intelligence and Natural Man.* New York: Basic Books.

Carey, S. & Diamond, R. (1980). "Maturational Determination of The Developmental Course of Face Encoding." *Biological Studies of Mental Process.* Ed. D. Caplan. Cambridge: MIT Press.

Chomsky, N. (1975). *Reflections on Language.* New York: Pantheon.

Cosmides, L. (1985). *Deduction or Darwinian Algorithms?: An Explanation of The Elusive Content Effect on The Wason Selection Task.* Doctoral Dissertation, Department of Psychology, Harvard University, on microfilm.

Cosmides, L. & Tooby, J. (1987). "From Evolution to Behavior: Evolutionary Psychology as the Missing Link." *The Latest on the Best: Essays on Evolution and Optimality.* Ed. John DuPré. Cambridge, MA: MIT Press.

Daly, M. & Wilson, M. (1980). "Discriminitive Parental Solicitude: A Biological Perspective." *Journal of Marriage and the Family* **42**, 277–288.

Daly, M. & Wilson, M. (1981). "Abuse and Neglect of Children in Evolutionary Perspective." *Natural Selection and Social Behavior.* Eds. R. D. Alexander & D. W. Tinkle. New York: Chiron Press.

Daly, M. & Wilson, M. (1982). "Homicide and Kinship." *American Anthropologist* **84**, 372–378.

Daly, M. Wilson, M. & Weghorst (1982). "Male Sexual Jealousy." *Ethology and Sociobiology* **3**, 11–27.

Hamilton, W. D. (1964). "The Genetical Evolution of Social Behavior." I and II. *Journal of Theoretical Biology* **7**, 1–52.

Marr, David (1982). *Vision: A Computational Investigation into The Human Representation and Processing of Visual Information.* San Francisco: W. H. Freeman.

Maynard Smith, J. (1964). "Group Selection and Kin Selection." *Nature* **201**, 1145–1147.

Sahlins, M. D. (1976). *The Use and Abuse of Biology.* Ann Arbor: University of Michigan Press.

Shepard, R. N. (1984). "Ecological Constraints on Internal Representation: Resonant Kinematics of Perceiving, Imagining, Thinking, and Dreaming." *Psychological Review* **91**, 417–447.

Shepard, R. N. (1981). "Psychophysical Complementarity." *Perceptual Organization.* Eds. M. Kubovy & J. R. Pomerantz. Hillsdale, NJ: Lawrence Erlbaum Associates, pp. 279–341.

Tooby, J. & Cosmides, L. (in press). *Evolution and Cognition.*

Tooby, J. & DeVore, I (1987). "The Reconstruction of Hominid Behavioral Evolution through Strategic Modeling." *The Evolution of Human Behavior: Primate Models*. New York: SUNY Press.

Trivers, R. L. (1971). "The Evolution of Reciprocal Altruisms." *Quarterly Review of Biology* **46**, 35–57.

Wanner, E. & Gleitman, L. (1982). *Language Acquisition: the State of the Art.* Cambridge: Cambridge University Press.

Wason, P. C. & Johnson-Laird, P. N. (1972). *Psychology of Reasoning: Structure and Content*. Cambridge, Mass.: Harvard University Press.

Williams, G. C. (1966). *Adaptation and Natural Selection*. Princeton: Princeton University Press.

RICHARD W. WRANGHAM
Department of Anthropology, University of Michigan, Ann Arbor

War in Evolutionary Perspective

Since the mid-1960's, inclusive fitness theory has revolutionized the study of animal behavior, and it now promises far-reaching changes for the social sciences (DeVore, this volume). A synthesis will not come easily, however. On the one hand, biologists tend to trivialize the complexities introduced by features such as language, culture, symbolism, ideology and intricate social networks. On the other hand, most social scientists have a strong aversion to reductionism even within their own fields, let alone when imported from the alien culture of biology. A shotgun marriage of biologists and social scientists is more likely to engender mutual hostility and deformed offspring than hybrid vigor.

Adding to the difficulties, the biological analysis of animal behavior, though surging forward, is still at a primitive stage. One of the areas with the firmest body of theory, for instance, is the study of sex ratios in maternal broods (Charnov, 1982). Elegant models predict different ratios under different conditions. In some cases the models work beautifully, but in others they fail. This will not surprise anyone familiar with the complexity of living systems. It stresses, however, that inclusive fitness theory is still feeling its way even within the mathematically tractable areas of biology. Sceptics in social science can therefore afford their doubts. Partly for this reason I want to discuss a topic where the biological component, while seeming in some ways to be unimportant, is nevertheless so striking that it cannot be ignored.

ANIMAL BEHAVIOR AND HUMAN WARFARE

Human intergroup aggression ranges from ambushes designed to club a member of a neighbouring band, to switches flipped to release a nuclear bomb; from half-a-dozen men facing each other with spears and warpaint, to hundreds of thousands maintaining a line with guns and aerial support; from an agreement round a campfire that enough is enough, to a dictator forcing soldiers into battle against their better judgment; from the effort to retrieve a kidnapped sister to suicidal anger at an insult to one's god. The cultural, technological and ideological components of war are so evidently strong that it is easy to dismiss a biological analysis as irrelevant (e.g., Montagu, 1968; Beer, 1981).

The biologist's general answer is that four billion years of intense natural selection must surely have shaped the human psyche in ways that allow us to understand aspects of the behavior of even computer-age humans. But in this case there is a more convincing and specific point, produced by animal field studies during the last 25 years. The social organization of thousands of animals is now known in considerable detail. Most animals live in open groups with fluid membership. Nevertheless there are hundreds of mammals and birds that form semi-closed groups, and in which long-term intergroup relationships are therefore found. These intergroup relationships are known well. In general they vary from benignly tolerant to intensely competitive at territorial borders. The striking and remarkable discovery of the last decade is that only two species other than humans have been found in which breeding males exhibit systematic stalking, raiding, wounding and killing of members of neighbouring groups. They are the chimpanzee (*Pan troglodytes*) and the gorilla (*Pan gorilla beringei*) (Wrangham, 1985). In both species a group may have periods of extended hostility with a particular neighbouring group and, in the only two long-term studies of chimpanzees, attacks by dominant against subordinate communities appeared responsible for the extinction of the latter.

Chimpanzees and gorillas are the species most closely related to humans, so close that it is still unclear which of the three species diverged earliest (Ciochon & Chiarelli, 1983). The fact that these three species share a pattern of intergroup aggression that is otherwise unknown speaks clearly for the importance of a biological component in human warfare. The divergences probably occurred between five and ten million years ago. The strong implication is that all three species have had extraordinarily aggressive intergroup relationships for the same length of time. Hence, not only has natural selection had the opportunity to shape psychological features underlying motivations in intergroup aggression, but also we have two closely-related species in which we can expect to be able to test simple aspects of theories of human aggression.

Warfare is only one of a huge array of human social relationships with visible origins in the animal kindom. Many others are more easily investigated because they are more widespread in animals. Alliances between kin, for example, or conflict between parents and offspring, occur in so many species that there are numerous opportunities for the testing and refinement of theories (Daly & Wilson, 1983; Trivers,

1985). It is not for the tractability of the problem that I focus on warfare, however. Understanding the ultimate causes of war is an important goal because of the remote possibility that an improved analysis will lead to better systems for preserving peace. Given that biology is in the process of developing a unified theory of animal behavior, that human behavior in general can be expected to be understood better as a result of biological theories, and that two of our closest evolutionary relatives show uniquely human patterns of intergroup aggression, there is a strong case for attempting to bring biology into the analysis of warfare. At present there are few efforts in this direction (but see Durham, 1976). Social scientists are needed who will invite biologists to work with them on the problem. This will only happen if they are persuaded of the value of biology.

THEORETICAL APPROACHES

The value of biology for an understanding of warfare is still a matter of faith, because no models yet account for the distribution and intensity of war or intergroup aggression. To a biologist the faith appears justified for two reasons. First, the cost-benefit analysis offered by inclusive fitness theory has a convincing theoretical rationale (natural selection theory) and a clear currency of measurement (genetic fitness). Second, it has achieved substantial success in explaining both species differences in social behavior and the dynamics of particular systems (Krebs & Davies, 1984; Trivers, 1985). As mentioned above, however, comparatively few tests have achieved mathematical precision. Biologists are therefore forced to be content (for the moment) with a level of analysis which is crude by comparison with harder sciences, even though it is sophisticated in relation to behavior theories of a few years ago.

The failure of social sciences to share biology's faith in itself is partly the result of an outmoded concept of biological theories. It is commonly suggested, for instance, that a biological analysis of human behavior implies a reactionary politics (Caplan, 1978). Evolutionary biologists, by this view, present a scenario of human behavior with little room for change because it relies on the assumption that behavior emerges from unmodifiable instincts. It is certainly true that animal behavior was once thought widely to be instinctual. Indeed the idea of an inborn aggressive drive was applied to humans and other animals only twenty years ago (Lorenz, 1966). But the interaction between inclusive fitness theory and ecology, together with field observations of animals, has led to a wholly new view of individual animals as strategists capable of modifying their actions in their own interests. According to this new concept, animals respond adaptively even to short-term changes in their environments (Emlen, 1976). Individuals in more intelligent species can respond adaptively to a wider range of novelties, whether in the social or nonsocial realm. This means that as the biological analysis of human nature becomes more sophisticated, it makes fewer abstract statements about the nature of society. Behavior emerges from the interaction between environment and individual. It is an adaptable system, not a fixed set of patterns.

This conclusion is relevant to another worry of social scientists about the implications of biological analysis, that biological explanations are inappropriate because, with the development of language (or other uniquely human traits), humans leave the evolutionary realm. Such a comment is partly fair. The predictions of inclusive fitness theory apply directly only to species which have had time to adapt to their environments. Hence, many aspects of human behavior doubtless fail to conform to evolutionarily based predictions. But this is far from saying that inclusive fitness theory is unhelpful. Even if the modern environment is too different from earlier environments to have allowed a fully appropriate human psychology to evolve, inclusive fitness theory still has important conclusions to offer about the proximate mechanisms by which individuals achieve their (possibly non-adaptive) goals.

For example, the simple evolutionary prediction is that more powerful individuals will use their power to achieve higher fitness. Yet we observe that the rate of population growth in the U.S.A. is much lower than in many poorer countries, or that within the U.S.A. the families of the rich are hardly larger, if at all, than those of the poor (Daly & Wilson, 1983). Two possibilities follow, both consistent with biological theory. First, power may no longer be correlated with fitness. If so, we expect that power is used in modern environments in a manner that would have been adaptive in previous environments. For instance it may be used to obtain an intervening variable that was formerly highly correlated with fitness, such as social status. Second, it is possible that power still is correlated with fitness, despite appearances from first-generation analysis. These alternatives are testable. Until they have been examined, no conclusions about their relative merits are legitimate. If either turns out to be valid, as an evolutionary biologist would expect, inclusive fitness theory will be useful for an analysis of contemporary behavior.

It will be relevant, for instance, whenever we need to understand human motivations. This is certainly an outstanding problem in the context of war, whether we think of leaders, soldiers, pacifists or observers. To illustrate the kinds of approaches which may attract interdisciplinary analysis, I consider briefly three recent ways in which biologists have been trying to understand the distribution of aggressive motivations. They are concerned primarily with aggression between individuals, but are easily extended to groups.

First, game theory has been modified to allow its application to evolving populations (Maynard Smith, 1982). Animals are envisaged as having a finite set of possible strategies which they use in interactions with each other. The best strategy set depends on what other individuals do. The key concept is that there exists an "evolutionary stable strategy" (or ESS), such that if all members of the population adopt it, no other strategy can invade. The ESS may be pure, in which case the individual behaves the same in all contexts. The interesting cases are those where the ESS is a mixed strategy, however, such that an individual adopts different strategies with different probabilities.

Game theory has been applied extensively to animal contests using simple assumptions. In the "Hawk-Dove-Retaliator" game, for example, three strategies are matched. "Hawk" invariably fights until injured or victorious; "Dove" displays but retreats if the opponent fights; "Retaliator" behaves like a Dove against a Dove, but as a Hawk against a Hawk. Depending on the pay-off in encounters between strategies, different evolutionarily stable states emerge. The ESS typically converges on a stable attractor point, and is therefore not necessarily affected by the initial frequency of strategies. However, it is strongly affected by the pay-offs to each strategy in different types of encounter. By showing how different kinds of behavior are favored depending on pay-offs in dyadic interactions, these models promise a sophisticated correlation of animal behavior with ecological conditions.

The second approach considers social relationships rather than social interactions (Popp & DeVore, 1979). The difference is that in social interactions (as analysed, for example, in most game theory models) individuals do not modify their behavior according to past experience with their opponent, whereas in social relationships they do. This is therefore a more realistic method, derived from the study of animals in complex social groups. It stresses that an anlysis of aggressive behavior must take account of costs and benefits not only to the opponents competing over a resource, but also to their future relationship with each other and with witnesses. Only then can one explain, for example, why opponents frequently reconcile with each other after an aggressive incident, or why intense competition may occur for status, even in the absence of an immediate resource (de Waal, 1982). The biological analysis of social relationships is in its infancy, but already it promises to allow easy bridges to the social sciences because it takes explicit account of the complexity of social networks.

The previous approaches are concerned with conflict, but not with totally unprovoked aggression. A third approach seeks to account for aggression even when neither resources nor status appear to be at stake. The classic area of investigation here is the study of infanticide. Although infants are sometimes killed for food or because they are competitors, an overwhelming mass of evidence shows that in many cases infanticide occurs because the death of the infant benefits the reproductive interests of the killer. The distribution of infanticide appears to be well correlated with variations in the vulnerability of infants and the risks and benefits to the killers (Hausfater & Hrdy, 1984). The implication from these studies is that natural selection favors unprovoked aggression provided that the target is sufficiently vulnerable, even when the benefits are not particularly high. Spontaneous aggressive motivations need not be merely the result of fear or pathology, however distasteful the idea.

These approaches indicate how biologists are looking for rules governing the frequency and context of animal aggression. They look valuable because despite the simplicity of their assumptions, they can generate realistic analyses of behavior, still rooted in inclusive fitness theory. The establishment of analytical principles will open the way to a synthesis of biological and social sciences if they succeed in showing the logic behind the evolution of aggressive motivation.

More than anything the biological approach stresses the rewards of working with a logically sound currency in cost-benefit analyses. There are many areas of social science, of course, that bring economic calculations to the analysis of war. Historians are often wrapped in strategic thought (Howard, 1983), as are many models in political science (Singer, 1980). Anthropology brings a more diverse approach. For instance, it emphasizes the importance of the relationship between warfare and social rules, but it also includes a variety of cost-benefit analyses (Fried et al., 1968; Otterbein, 1970). The currency in these models is a matter of debate, however, so that elaborate theories are easily criticized. Strategic thinking in biology is often simplistic by comparison, but it has the great merit over the social sciences of an ultimate theoretical foundation. The result is that model-building can proceed from a safe base.

REFERENCES

Beer, F. A. (1981). *Peace Against War: The Ecology of International Violence.* San Francisco: W. H. Freeman.

Caplan, A. L. (1978). *The Sociobiology Debate.* New York: Harper & Row.

Charnov, E. L. (1982). *The Theory of Sex Allocation.* Princeton University Press.

Ciochon, R. L. & R. S. Corrucini, eds. (1983). *New Interpretations of Ape and Human Ancestry.* New York: Plenum Press.

Daly, M. & M. Wilson (1983). *Sex, Evolution and Behavior.* Boston: Willard Grant Press.

Durham, W. H. (1976). *Q. Rev. Biol.*, **51**, 385–415.

Emlen, S. T. (1976). *Science*, **192**, 736–738.

Fried, M., M. Harris & R. Murphy, eds. (1968). *War: The Anthropology of Armed Conflict and Aggression.* New York: Doubleday.

Hausfater, G. & S. B. Hrdy, eds. (1984). *Infanticide.* Hawthorne, New York: Aldine.

Howard, M. (1983). *The Causes of Wars.* London: Unwin Paperbacks.

Krebs, J. R. & N. B. Davies, eds. (1984). *Behavioural Ecology.* Sunderland, Mass.: Sinauer.

Lorenz, K. (1966). *On Aggression.* New York: Harcourt, Brace & World.

Maynard Smith, J. (1982). *Evolution and the Theory of Games.* Cambridge University Press.

Montagu, M. F. A., ed. (1968). *Man and Aggression.* Oxford University Press.

Otterbein, K. (1970). *The Evolution of War.* New Haven: HRAF Press.

Popp, J. L. & I. DeVore (1979). In: *The Great Apes.* Ed. D. A. Hamburg & E. R. McCown. Menlo Park: Benjamin/Cummings.

Trivers, R. L. (1985). *Social Evolution.* Menlo Park: Benajmin/Cummings.

de Waal, F. (1982). *Chimpanzee Politics.* New York: Harper & Row.

Wrangham, R. W. (1987). In: *Primate Models of Hominid Evolution.* Ed. W. Kinzey. New York: SUNY Press.

DOUGLAS SCHWARTZ
School of American Research, Santa Fe, New Mexico 87501

The Relationship of Modern Archeology to Other Disciplines

SUMMARY OF REMARKS

Archeology requires interaction with other disciplines as soon as it goes beyond the simple description of the surviving remnants of past cultures. The archeologist must attempt to place his descriptions in historical sequence; to view past cultures as interacting, open systems; and to identify major changes in these systems and to specify their causes. He must then extract cross-cultural regularities that may suggest laws of human behavior and of cultural evolution. The further he moves from simple description to interpretation, the more dependent he is on other disciplines.

There are three big questions that archeologists must ask when they turn to interpretation. We can formulate these questions as follows:

1. When did the essence of humanity emerge in nonhuman primates and what caused the change?
2. Why did agriculture and fixed settlements replace nomadic hunting and gathering?
3. What forces triggered the development of cultural complexity, including craft specialization, the development of elites, and the emergence of power based on economic, religious, and other forces?

Work on the first question is going on mainly in Africa. The southwestern United States is a tremendous laboratory to study the second question and, to an increasing extent, the third. There are intriguing examples in the Southwest of the development of major centers of economic complexity, which eventually disappeared due to drought or for other reasons. The movement toward greater complexity was truncated in 1500 with the arrival of the Spanish.

It is not enough to appeal only to the physicists and chemists for help. We must also look to the humanities to understand the motivations of past cultures. So archeology is an excellent example of an activity which integrates contributions from a whole series of interrelated disciplines. This is, of course, what the Santa Fe Institute hopes to do on an even larger scale.

ANTHONY TURKEVICH
The Enrico Fermi Institute and the Department of Chemistry, The University of Chicago,
5640 Ellis Avenue, Chicago, Il 60637

Reconstructing the Past through Chemistry

The purpose of these remarks is to touch on a few topics that illustrate the interaction of chemistry with other disciplines. These topics can serve as examples of the types of activities that might be undertaken by the Santa Fe Institute. Among the many that might be used, the focus here will be on "Reconstructing the Past through Chemistry."

Reconstruction of the past is an interdisciplinary activity that involves physicists, chemists, geologists, and paleontologists, among others. It is of interest to a wide public and there have been spectacular developments recently. At the same time, such studies do not have a home in any traditional academic department. They are carried on, and to a certain extent, effectively, in various disciplinary departments: geology, physics, chemistry, and archeology. Each of these tends to have a particular slant governed by its disciplinary environment. Thus, providing a home that does not have such biases is a fruitful area to explore for the Santa Fe Institute. Although some previous talks have touched on this area, the present remarks should not unduly overlap those. Two of the examples that will be covered will be from the physical world; the last examples will be from archeology.

The past has extended over such a long time, that only infinitesimally small parts can be studied in any detail. Only a year after the start of our universe in the Big Bang, when electrons started to recombine with protons, alpha particles and a smattering of heavier elements, chemical reactions began. Possibly some of the few hydrogen molecules formed at that time have survived to this day. Some of the more complicated molecules in interstellar space are certainly older than our solar system. Thus, radioastronomers studying such molecules are doing a chemical reconstruction of the past as they address questions of the formation processes leading to these molecules.

Turning to the more recent past, there have been significant discoveries in recent years on the state of the solar system at the time that it was condensing from a gaseous solar nebula, or agglomerating from a cloud of dust and gas. Of the material available to study the state of the nebula at that time, terrestrial samples are almost worthless since they have been greatly distorted by the geological processes that have gone on through intervening time. Much more useful are meteorites and samples of the moon. Finally, there is the dust which can be collected from the stratosphere, dust which may well be the debris of comets which, in some theories, have survived from the time of the formation of the solar system. Such stratospheric dust samples are too small for extensive studies. It is the other samples of extra-terrestrial material (lunar and meteoritic) that are being intensively studied for their chemical and isotopic composition in various laboratories. The relative abundance of the chemical elements that are present, the mineralogy, the microscopic physical structure, all provide evidence on the state of the early solar system.

Perhaps the most intriguing information comes from accurate measurements of the isotopic composition of various chemical elements in samples of extra-terrestrial matter. In the case of several elements, for example, neon, magnesium, and oxygen, such measurements indicate that the material of our solar system was not produced in a single star, but had its origins in several places in the galactic nebula.

The case of the isotopic composition of oxygen has been studied in some detail. Oxygen has three stable isotopes: oxygen 16, oxygen 17, and oxygen 18. The relative abundances of these three isotopes of oxygen in various samples have been carefully measured by Professor Robert Clayton at the University of Chicago.[1] Oxygen is a most pervasive element. It is the most abundant element in the condensed matter of most samples and yet this element retains evidence of conditions in the early solar system.

Figure 1 shows the nature of this evidence. The figure is a plot of the ratio of oxygen 17 to 16 [$\delta^{17}O(SMOW)$] versus the ratio of oxygen 18 to oxygen 16 [$\delta^{18}O(SMOW)$]. Both coordinates are in parts-per-thousand deviation from the values in a reference sample. The total range is small—but unmistakable. The isotopic composition of oxygen is *not* exactly the same in all samples. The values from all *terrestrial* samples lie on a line with slope one-half and have an error which is not much larger than the width of the line. Samples of oxygen from terrestrial silicate rocks or snow or biological material provide values that lie along this line.

One may ask why there is any spread at all in the values from these samples. Why isn't the isotopic composition of all terrestrial samples *exactly* the same? On this fine scale, there is a spread because there are mass effects on physical and chemical equilibria and on the rates of these processes. It can be shown that all these should produce effects twice as large for oxygen 18 as for oxygen 17. This explains why all terrestrial samples lie on a line with slope one-half in Figure 1.

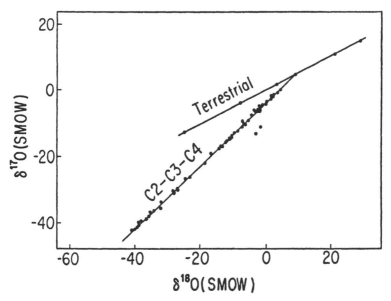

FIGURE 1   Figure 1. Variation of oxygen isotyope ratios in terrestrial and certain meteoritic (C-2, C-3, C-4) materials. From Clayton[1].

When Clayton examined some extra-terrestrial samples, he found quite different results. The data from some meteorites, such as the C-1 chondrites, lie pretty close the the terrestrial line. But the data from several classes of meteorites, C-2, C-3, and C-4 carbonaceous chondrites, are off the line—much beyond experimental error. Such meteorites, from their morphology and mineralogy, appear to have been subjected to less geological alteration than any other meteorites. They look as if they had been gathered together and merely packed together from the pre-solar material. They certainly haven't been melted or subjected to excessively high pressures and some even have water in them, so that they could never have been at a high temperature. They are, at present, the best examples of early solar system material. Not only do the data from such meteorites in Figure 1 lie off the terrestrial line, but they have a quite different slope. Instead of a slope of one half on this scale, which is that predicted theoretically for mass effects, the data from these meteorites lie on a line of unit slope.

Such a slope could be explained if, in addition to oxygen of ordinary isotopic composition, there had been added *some pure oxygen 16*. This would imply that these particular meteorites came from a region of the solar system that had a different amount of oxygen 16 injected into it than the earth and moon. Therefore, the preagglomerated solar system was not homogeneous isotopically. And if it was not homogeneous isotopically, then it probably was not homogeneous chemically, although so far no concrete evidence of this has been found. This undoubtedly is because techniques for establishing chemical heterogeneity are not as sensitive as those for establishing isotopic heterogeneity.

Recently questions have been raised whether photochemical processes in the early solar nebula could have produced the small isotopic changes that are observed in the samples being discussed. However, the clustering of the anomalous oxygen isotopic ratios about the line with unit slope, and the magnitude of the effects produced, appear to be inconsistent with possible photochemical processes, as is the presence in many of the same samples of isotopic anomalies of other elements.

The second example of the role of chemistry in reconstructing the past involves not only chemistry, but geology, paleontology, and meteorology, but it all started with chemical analyses. It, thus, is a prime example of an interdisciplinary activity of the type proposed for this Institute. This is the discovery by the Alvarez's, father and son, and their group at Berkeley,[2] that certain geological strata have elevated concentrations of the rare element iridium. This enrichment, at the boundary between the sediments deposited in the Cretaceous and Tertiary periods, about 65 million years ago, is world wide. It has now been established at many sites on at least four different continents.

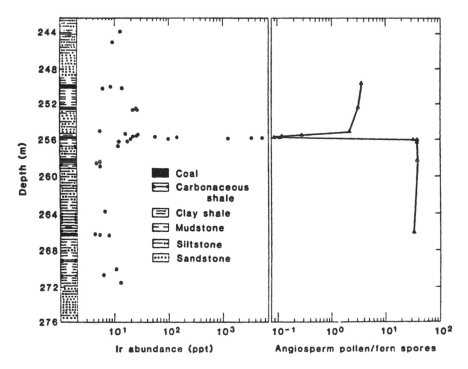

FIGURE 2   Figure 2. Iridium abundance (in parts per trillion) as a function of depth (in meters) near the Tertiary-Cretaceous Boundary. Also shown are the concentrations of pollen spores (from C. J. Orth, et al., ref. 3).

One of the sites where extensive work has been done is only about 100 miles north of here, near Trinidad, Colorado. A group at the Los Alamos National Laboratory, led by Dr. Carl Orth,[3] has obtained some of the most striking data. Figure 2 from their work shows the iridium abundance, in parts per trillion, plotted horizontally on a logarithmic scale as a function of depth in the rock, in meters, on the vertical axis. At the Tertiary-Cretaceous Boundary, the iridium abundance rises to values hundreds of times higher than in strata above or below the boundary. Paleontogical studies have shown that at the same time, many species of plants and animals disappeared. This is illustrated on the right side of Figure 2. It has been proposed that the enhanced iridium in this world-wide deposition is due to the collision with the earth of an asteroid or comet. Such objects are expected to have iridium contents more than thousands of times higher than terrestrial surface rocks. The injection into the high atmosphere of the debris from such a cataclysmic event could have spread the iridium world wide and, it is theorized, produced the extinction of life forms that appear to have occurred at this time. This discovery has also sparked speculation that a significant number of large *nuclear* explosions, by injecting dust into the atmosphere, could produce climatic effects large enough to affect the ability of parts of the planet to sustain life.

The interpretation of the elevated iridum contents of certain geological strata as being due to extra-terrestrial impacts is not accepted by everybody. An alternative explanation suggests that enhanced volcanic activity over a period of 10,000 to 100,000 years[4] is the cause of the enhanced noble metal and other elemental abundances, as well as being responsible for the extinctions of species that appear to have occurred at the same time. Whichever explanation turns out to be true, chemical analysis for one of the rarest elements has impacted on studies of geology, evolution, meteorology and even on contemporary considerations of military policy.

The third example involves the use of various types of chemical analyses for establishing the times and nature of artifacts left over from previous human activity. There are many facets to this and it is a thriving occupation. Again, it is an activity that is inherently multidisciplinary. A very few illustrative examples will be noted here.

A typical situation is to take shards of pottery and try to establish either their origin or the period when the pottery was made or used. Using modern analytical techniques, the relative amounts of 15 to 30 chemical elements can be determined accurately on very small samples. In this way the source of pottery used in the Eastern Mediterranean 3000 years ago can be identified as being either from Crete or from mainland Greece by the distinctly lower amounts of chromium and nickel in the pottery from the mainland.[5] The trends in the amounts of the other elements support this interpretation.

Another example is the history of the introduction of European pottery-making techniques into the Americas. How long did the Spanish in the Americas import their pottery and when did they start making their own? There has been an extensive study of remains of pottery used by the Spaniards in the early days of their American stay.[6] In the case of the settlements in Venezuela and the Dominican Republic, all the pottery was imported for a long time from the homeland in Spain, in fact from one place near Cadiz. On the other hand, in Mexico, within 50 years after the conquest, the Spanish were making their own pottery out of local clay with the same kind of surface glaze and decorations that they were used to having back in Spain.

Finally, an analytical technique of a different type may be mentioned. This involves the use of racemization of optically active molecules. Our bodies, and all living material that we know, are composed of only *levo*-amino acids (molecules that have only one out of the two possible geometric arrangements of atoms about a center of asymmetry in the molecule). On the other hand, at a given temperature, levo-amino acids gradually racemize, converting slowly into an equal mixture of levo and dextro molecules. This conversion is happening in our bodies all the time, fortunately slowly enough so that there are no ill effects, since the dextro forms cannot be utilized in body chemistry. The extent of this racemization can be used for various chronological purposes. Since the rate is very temperature sensitive, the rate of racemization of the amino acids of a human in the arctic region will be drastically reduced when death occurs and the body is placed in the low-temperature earth. The amount of racemization in human remains in arctic regions has been used to estimate the age at which the person dies.[7]

On the other hand, in more temperate regions, the rate of racemization is signficant even after burial. For certain amino acids the half-life for racemization under these conditions is about 15,000 years. Thus, the technique has potential for providing chronological information going even further back than carbon-14, with its half-life of 5730 years. Although this technique is in its infancy and its limitations have to be explored,[8] it also has the potential for making use of many more amino acids than have been studied so far.

These are examples of the use of chemistry to study and to reconstruct the past. They are clearly only a minute fraction of the work that has been done, and only an indication of the possibilities for the future. The location here near Santa Fe appears to be ideal for such studies. There are museums, interest in anthropology, much geological activity in the Southwest, and superb facilities for chemical analyses at the Sandia and Los Alamos National Laboratories. Thus, an Institute with these interests would be building on, and expanding, a local base, as well as making use of, through modern communications, contacts with work in this area in the rest of the country and even the world.

REFERENCES

1. R. N. Clayton, *Philosophical Transactions Royal Society of London*, A **303**, 339 (1981); R. N. Clayton and T. K. Mayeda, *Abstracts of Eighth Lunar Science Conference*, Houston, Texas, **Part 1**, p. 194 (1977). See also Clayton data in C. T. Pillinger, *Geochimica et Cosmochemica Acta.* **48**, 2739 (1984).
2. L. W. Alvarez, W. Alvarez, F. Asaro, H. V. Michel, *Science* **208**, 1095 (1980).
3. C. J. Orth, J. S. Gilmore, J. D. Knight, C. L. Pillmore, R. H. Tschudy, and J. E. Fasset, *Science* **214**, 1341 (1981).
4. C. B. Officer and C. L. Drake, *Science* **227**, 1161 (1985).
5. G. Harbottle et al., *Archeometry* **11**, 21 (1969).
6. J. C. Olin, G. Harbottle, and E. V. Sayre, "Archeological Chemistry, II," Ed. Giles F. Carter, *ACS Advances in Chemistry Series* **171**, 200 (1978).
7. P. M. Helfman and J. L. Bada, *Nature* **262**, 279 (1976).
8. P. M. Masters and J. L. Bada, "Archeological Chemistry, II," Ed. Giles F. Carter, *ACS Advances in Chemistry Series* **17**, 117 (1978).

JEROME L. SINGER
Director, Clinical Psychology, Yale University, New Haven, CT 06520-7447

The Conscious and Unconscious Stream of Thought

In presenting an overview on the current state of psychological research on the stream of conscious and unconscious thought, I am sensitive to the fact that, in a way, my auditors or readers are already experts on the subject. Each of you knows the nature of your ongoing thoughts, fantasies, memories, your verbal glosses on the passing scene—indeed, you know that you process a vast amount of "internal" information to which the psychologist is not privy unless you choose to reveal it. Indeed, you might try a thought experiment and take note of the many times your attention wanders away from my content in the direction of an awareness of hunger pangs, perhaps some thoughts about your evening plans or, even more remotely from your task as a reader or auditor, to an extended romantic or exotic fantasy. A major challenge to modern psychology, psychiatry or behavioral science generally is to construct models of the dynamic flow of human responses which incorporate the interplay between the publically observable movements and speeches of people and the fact that each individual carries on some complex mixture of private conscious thought and, indeed, very likely some form of unconscious mentation during every type of social or solitary action.

A FRAMEWORK FOR STUDYING ONGOING THOUGHT

1. THE PSYCHOANALYTIC METHOD

Sigmund Freud and William James met in the United States at Clark University in 1909. The brief encounter of the two great pioneers of the scientific study of the flow of human thought processes symbolizes the task which modern students of consciousness must now confront. William James, using introspection and clinical observation (Taylor, 1983), described the critical properties of ongoing conscious thought as a basic dimension of human psychology in his classic textbook (James, 1890/1952). Sigmund Freud used the characteristics of the thought stream as verbalized in the free associations of his patients to identify through blockages of verbalizations, diversions in sequence, and momentary forgettings, the operation of a set of thought activities that were unconscious or preconscious. Much of the modern psychoanalytic theorizing about the ways in which presumably unconscious wishes, fantasies, conflicts or interpersonal attitudes (transferences and object-representations) influence adult behavior continues to be derived from anecdotal accounts of psychoanalysts who are assumed to be well-trained to make observations and to draw inferences from samples of the free associative thought. Indeed, to the extent that one can assert that psychoanalysis meets the various criteria of eliminative inductionism and remains a viable scientific method for investigating the possibility of unconscious influences upon the public personality of an individual (Edelson, 1984), one must confront the method's reliance upon ongoing associative thought as its information base.

An example from recent research which demonstrates a systematic quantitative approach to using the free associations of patients in psychoanalysis to test an aspect of Freud's theory of repression, the defensive nature of momentary forgetting, has been provided by the ingenious work of Luborsky (1977). Figure 1 shows the average relationship for ten patients between their speech patterns, sudden forgetting of thought or other material, and subsequent flow of speech following the episode of forgetfulness. By examining the tape-recorded transcripts of actual therapy sessions, Luborsky was able to show that as patients touched on difficult topics—usually their relationship with the therapist (the transference)—there was an increase in hesitations, exclamations of uncertainty, and other signs of cognitive disturbance (see left half of graph). After the instance of forgetting, the speech disturbances are reduced (right half of graph), suggesting that the defense of repression has served temporarily to reduce anxiety. Studies of this type provide some of the best evidence so far available to back up Freud's original observations of how defenses worked in the psychoanalytic session.

While the hypotheses about unconscious thought activity derived from psychoanalysis reflect sweeping insights, we may have underemphasized the valuable harvest to be gleaned from studying the domain of normal conscious thought. These waking interior monologues, reminiscences, mental rehearsals of future acts, the speculative forays into possible and impossible futures we call daydreams or fantasies, all are part of the ongoing flow of consciousness first identified formally by

William James. Curiously, Freud and many subsequent psychoanalytic theorists have paid surprisingly little attention to the structural characteristics of natural-occurring associative thought despite their dependence on the content of such material for drawing inferences about unconscious mechanisms. My own hypothesis which someday I hope to elaborate is that Freud's Victorian prudishness led him to cast the childish, trivial, slimy, salacious, self-serving and hostile qualities of ordinary conscious thought down to the limbo or hell of an unconscious mind. Rather than confront the full absurdity of much of our ongoing consciousness, he emphasized the secondary process or logical sequential processes of ego-oriented consciousness and studied the primary processes as manifestations from the nether regions, discernible in occasional peremptory ideational upsurges, transference fantasies and, especially, in night dreams.

With the emergence of the cognitive movement in the behavioral sciences from about 1960 on, we see a paradigm shift towards a view of the human being as an information-seeking, information-processing organism rather than as a hydraulic energy machine, a creature endlessly struggling to contain the pressures from sexual or appetitive drives, a view apparent from the writings of psychoanalysts and learning theorists of the 1930s, '40s and '50s. Yet even the cognitive movement with its focus on the active sequence of information-organization is somewhat uncomfortable with the problem of the natural stream of thought. Most cognitive research assigns people circumscribed, well-defined problems to solve, whether in the form of the indentification of rapidly presented letters, shapes, pictures, etc. Even the

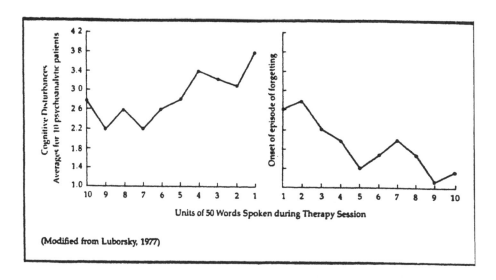

(Modified from Luborsky, 1977)

FIGURE 1

revival of interest in private imagery has chiefly emphasized images as direct duplication of objective, external stimuli as in the experiments of Segal (1971), Shepard (1978) or Kosslyn (1981). Yet much natural-occurring imagery is more dynamic and fluid than the well-controlled metal cube rotations of Shepard and, indeed, it is probably much more about people in relationships or about buildings, shops or nature scenes than the geometric shapes we can manipulate so easily in the laboratory. In a sense, a painting such as Picasso's *Guernica* with its fragmented bodies, distorted horses and emotional impact captures the *memory images* of a spectator of the village bombing better than would a moving picture of the scene. Our great artists and writers have pointed the way for us in describing the role of conscious thought in the human condition. We now must move toward meeting that challenge by developing method and theory that make possible a fuller description of the functioning organisms as one that processes not only environmentally presented information about physical objects and people, but that also processes and reshapes a continuing flow of stimulation generated from one's own long-term memory system.

2. A COGNITIVE-AFFECTIVE PERSPECTIVE

It has become increasingly clear to cognitive psychologists that our ways of knowing the world are intrinsically bound up with our ways of feeling or, indeed, our moral and aesthetic evaluations (Rychlak, 1977, 1981; Tomkins, 1962-1963; Zajonc, 1980). Philosopher Robert Neville's "reconstruction of thinking" points to the centrality of some inherent valuation process in all knowing and certainly in imagination (Neville, 1981). Significant advances have been made in the past decade in empirical studies of the specific emotions with which we seem "wired." Excitement-interest and joy are positive emotions that, when invoked, are usually positively reinforcing. Fear-terror, distress-sadness, anger, and shame-guilt-humiliation are negative affects, generally serving as punishing experiences (Izard, 1977; Singer, 1974; Tomkins, 1962-1963).

Tomkins' proposal is that humans are inherently motivated by four implications of the positive and negative emotions: we maximize experiences we expect to generate positive affect and minimize the likelihood of experiencing negative affect; we experience and express emotions as fully as possible; and, finally, we control emotions as it becomes adaptively necessary. Since space limits a detailed exploration of the emotions, I will point here chiefly to their close link with the cognitive system and with the information-processing sequence. In effect, in studying the private personality, we need to recognize that we can be startled and intrigued by our own thoughts, that waking as well as nocturnal fantasies can evoke the fear or terror we associate with nightmares, that recurrent fantasies of betrayal or humiliation may have important bodily feedback implications, even if (or sometimes because) they are never translated into overt action. The quiet, "nonemotional" scholar can react with private experiences of intense joy to a humorous passage in one of Aristophanes' plays or with intense excitement at the realization of the relationship between two previously obscure readings of an ancient text. The hypertensive adult has been shown to be characterized specifically by recurrent aggressive daydreams (Crits-Christoph, 1984).

A key concept in the paradigm shift from an S-R to a cognitive perspective in psychology is the notion of a temporally extended, if very rapid, "sequence" in information processing. The close tie between information processing on the one hand and emotional experience on the other pointed to by Tomkins (1962-1963), Izard (1977), Mandler (1975), McClelland (1961), and Singer (1973,1974) has also greatly expanded our ability to relate motivation to cognition.

Most cognitive theories tend to emphasize consciousness as a feature of the private personality. They do not preclude, however, the possibility that many of our plans and anticipations may have become so automatic that they unroll too rapidly for us to notice them in the flurry of events. Thus, when we first learn to drive, we must consciously think of each step to be taken: "depress the clutch, shift the gear, gradually release the clutch, slightly feed the gas by the gas pedal." Once we have carried out a sequence like this often enough, we can engage in the complex motor and perceptual acts necessary for driving a car, and, simultaneously, talk, think of other events, listen to music, or observe the scenery. Langer's research on mindlessness, or overlearned mental action sequences is relevant here (Langer, 1983). Recently, Meichenbaum and Gilmore (1984) have developed further the viewpoint that unconscious processes reflect well-established or overlearned constructs, schemas, or metacognitions (e.g., rules of memory retrieval and various biasing rules about material accepting or threatening of self-beliefs), a position similàr to Tomkin's (1979) theory of nuclear scenes and scripts.

Cognitive theories often make the assumption that private experiences such as conscious thoughts, fantasies, or images provide an alternative environment to the continuous processing of material from the external world (Singer, 1974). Thoughts may be reshaped and reorganized and further acted upon by further thought in much the same way as our experience is modified by new inputs from the physical or social environment. Thus, there is a constant restructuring of material in the memory system; memory is never simply a process of passive storage.

Cognitive theories also assume that some attitudes, beliefs, or patterns of information are more central or self-oriented than others, and, therefore, are more likely to evoke complex affective responses. The self can be regarded as an object of cognition or as part of perceived experience rather than as an agent. Because our most personal schemata are associated both with a long background of memories from childhood and with our most recent experiences, they are linked to the most complex network of related images, memories, and anticipations. Novel material that does not fit in with beliefs or expectations will generate a sense of incongruity. In the face of persisting incongruity, an experience will evoke great intensities of distress or anger.

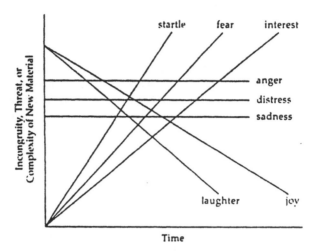

FIGURE 2 As the graph shows, a person's emotional reaction to a stimulus depends on the suddenness and incongruity of new information and the time it takes for incongruity to be reduced. Note that laughter and the positive emotion of joy are aroused when incongruity or threat is relatively quickly reduced, but that if high levels of incongruity persist, the negative emotions of anger, distress, or sadness may emerge. (Based on Tomkins, 1962)

INCONGRUITY AND THE AROUSAL OF SPECIFIC EMOTIONS. Tomkins (1962) has proposed an ingenious theory linking the sequence of arousal of specific emotions such as fear-terror, anger, sadness or joy to the suddenness of one's confrontation with complex or difficult-to-assimilate information and with the persistence of this incongruity over time. Figure 2 demonstrates the model. Let us consider, as a kind of thought experiment, the following example:

Let us suppose you happen to be home alone early one autumn evening. The doorbell rings. As you move to answer the door, the possibilities about who might be there quickly flit through your consciousness. It could be a friend or relative who occasionally drops in on an evening. A more remote possibility might be a magazine salesman, because you have heard from others that one has been around recently in the evening. In effect, then, even as you move toward the door, you are already drawing on your own background of memories—what Miller, Galanter, and Pribram as well as Tomkins would call your *image*; you are establishing some anticipation which can then be verified when you actually open the door.

There before your open door stands a gorilla! Your immediate reaction almost certainly would be to show the *startle reflex*. Your eyes blink, your arms are thrust up and back, your body is bent forward. Within a split second, you open your eyes and again see the gorilla and become overwhelmed with fear. In effect, you are confronting a stimulus that cannot be *matched* to any of your anticipated plans;

this produces a high level of incongruity or cognitively unassimilable material with an associated high level of density of neural firing within a very short time. The emotion of fear or terror is, thus, evoked by the situation.

Suddenly the gorilla says in a rather child-like voice: "Trick or treat!" Now you remember this is Halloween and, in an instant, you can make a mental match with a well-established custom, although not one you had been prepared for just at this moment. There is a sudden reduction in the novelty and complexity of the situation and you show the affect of joy, in a burst of relieved laughter.

Let us suppose for a moment that it was a real gorilla! This incongruity and threat of the situation persist at a high level and you cannot make any sense of it. The animal starts to advance into the house and you experience terror, then rage and anger at this intrusion. It forces its way in and you retreat back into the house and try to stem its advance, angrily throwing things at it while trying to find a source of escape. You are now a prisoner of the gorilla. It clomps around the house, knocking over furniture, breaking glasses, eating the fruit you had in a bowl, and you find yourself alternating between anger and despair. You experience a little more familiarity with the situation but still are helpless. With familiarity you are more likely to experience a somewhat lower level of persisting incongruity. This will lead to the affect of distress and sadness.

While my example, concocted some years ago, may seem unrealistic, a recent newspaper report described the case of eight chimpanzees who escaped from a traveling circus in West Germany. They created considerable distress and confusion by knocking at doors or appearing at the windows of local homeowners! So my fictional example may not be so far-fetched after all. One can surmise the emotions of the people who opened their doors to these straying apes.

SOURCES OF STIMULATION AND THE ONGOING THOUGHT PROCESS. To summarize my general point of view, the human being is confronted regularly by two major sources of stimulation, the complex physical and social characteristics of the surrounding environment which make demands for "channel space" on one's sensory system and an alternative, competitive set of stimuli generated by the brain that may also impact the sensory system although with somewhat less urgency when one is in the highly activated and aroused condition of wakefulness. A third source of stimulation, weaker in demand for conscious processing if often no less important, is the signalling system from the ongoing machinery of our bodies, a system of great importance in health but not yet well-enough researched and, certainly, except under great pain or fatigue, often ignored. What I would like to suggest is that as far as we can tell most people are carrying some kind of ongoing interior monologue, a kind of gloss on the immediately occurring events as well as engaging in associations of these events. Under circumstances in which the external stimulus field involves great redundancy or sufficient familiarity so that one can draw on automatized cognitive and motor processes, one may become aware of the continuing array of memories or fantasies unrelated to the immediate environment. Since, as I will argue below, much of our stress of thought is made up of unfinished intentions or longstanding as well as current concerns, the attention to such stimulation often provokes negative emotions of fear, sadness, shame-guilt or anger and has a generally mildly aversive quality. Thus, we often prefer to put on the radio or television, do crossword puzzles, or, if in an elevator with a stranger, talk about the weather, rather than stay with our thought sequences. Attention to self-generated stimulation does seem to involve at least temporarily a shift to a "different place" and the use of the same sensory systems, sometimes in parallel, sometimes in sequential fashion (Antrobus, Singer, Goldstein, & Fortgang, 1970; Singer, Greenberg, & Antrobus, 1971). The complex interaction of both hemispheres of the brain necessary for such a mixture of sequential thought and automatic verbal-chain or intended action-sequence processing (left hemisphere) and for the more parallel, global, novelty-seeking and perceptual orientation (right hemisphere) has been documented in an impressive review by Tucker and Williamson (1984). I will, however, focus the balance of this paper on a series of methods that have emerged for providing systematic data on ongoing thought and will not further address the presumed brain mechanisms that may underlie the recurrent generation of stored "material" that provides us with a phenomenal but very "real" experience of a stream of consciousness.

EXPERIMENTAL LABORATORY STUDIES OF ONGOING THOUGHT

1. STIMULUS INDEPENDENT THOUGHT IN SIGNAL DETECTION STUDIES

Beginning in 1960, John Antrobus and I developed a series of experiments designed to determine if we could in some way tap into ongoing thought. Our intention in effect was to capture the daydream or fantasy as it occurred, or come as close to doing so as possible. The model grew out of the vigilance and signal-detection studies developed in World War II to study how individuals could adjust to tasks that required considerable attention under monotonous conditions or environments of minimal complexity and stimulation.

In this model, the subject, in effect, has different degrees of demand made upon him or her for processing externally derived information under conditions of reasonably high motivation. Since the amount of external stimulation can be controlled, it remains to be determined by the study to what extent individuals will shift their attention from processing external cues in order to earn money by accurate signal detections, toward the processing of material that is generated by the presumably ongoing activity of the brain. Our attempt was to determine whether we could ascertain the conditions under which individuals, even with high motivation for external signal-processing, would still show evidence that they were carrying on task-irrelevant thought or, in the term we have been using more recently, stimulus-independent mentation (SIM).

If, while detecting signals, an individual was interrupted periodically, say, every fifteen seconds, and questioned about whether any stimulus-independent or task-irrelevant thoughts occurred, a "yes" response would be scored as SIM or TITR

(task-irrelevant thought response). By establishing in advance a common definition between subject and experimenter as to what constituted such task-irrelevant
thought, one could have at least some reasonable assurance that reports were more
or less in keeping with the operational definition established. Thus, a thought that
went something like the following, "Is that tone louder than the one before it? It
sounded like it was," would be considered stimulus-dependent or task-relevant and
would elicit a "no" response even though it was, indeed, a thought. A response
such as "I've got to remember about picking up the car keys for my Saturday night
date" would, of course, be scored as stimulus-independent mentation. A thought
about the experimenter in the next room ("Are they trying to drive me crazy?"),
even though in some degree generated by the circumstances in which the subject
found himself, was nevertheless scored as SIM because it was not directly relevant
to the processing of the signal that was defined for the subject as his or her main
task.

By keeping the subjects in booths for a fairly lengthy time and obtaining reports
of the occurrence of stimulus-independent thought after each 15-second period of
signal detection, it was possible to build up rather extensive information on the frequency of occurrence of SIM, their relationship to the speed of signal presentation,
the complexity of the task, and to other characteristics of the subject's psychological situation. Indeed, as Antrobus (1968) showed, it was possible to generate a
fairly precise mathematical function of the relationship of a stimulus-independent
thought to the information load confronted by the subject in the course of ongoing
processing.

By using periodic inquiries for content as well as for presence or absence of
SIM, it was possible to examine the range and type of content available and to
score this material along dimensions similar to those also used for night-dream
research, e.g., vividness of imagery, modality of imagery, degree of personal content
versus impersonal content, future or past references, etc. The alternative method of
establishing content was to make use of continuous free association by the subject
during a vigilance task (Antrobus & Singer, 1964).

A number of generalizations have emerged out of the signal-detection experiments. It was possible to indicate that stimulus-independent thought could be
reduced significantly if the amount of reward paid subjects or the complexity of
the task was systematically increased. As a matter of fact, although significant
reductions did occur, it turned out to be difficult to reduce reports of stimulus-
independent thought to zero unless signals came at such irregular intervals that
subjects could not apparently learn to pace themselves. While this would suggest
that the general pattern of dealing with stimulus-independent thought involves a
sequential style, there has been evidence in a study by Antrobus, Singer, Goldstein,
and Fortgant (1970) that, under certain circumstances, it is possible to demonstrate
parallel processing, that is, reports of stimulus-independent thought occurring even
as the subject was accurately processing signals.

When new, potentially personally-relevant information is presented to the subjects just prior to a signal-detection "watch," there is a greater likelihood of an

increase in stimulus-independent thought. Errors, however, may not necessarily increase for some time. It is as if, in many instances for tasks of this kind, subjects are not using their full channel capacity for processing private as well as external cues.

The signal-detection method for tapping in on ongoing thought presents some elegant opportunities for measuring more precisely what the odds are that any task-irrelevant mentation will take place at all. Fein and Antrobus (1977) were able to demonstrate that even though a trial of signal detections was increased from, say, one minute to two minutes (with signals coming every second, this would mean from perhaps 60 to 120 detections required of the subject), the relative frequency of reports of stimulus-independent mentation was capable of being described by a Poisson distribution once the subject made an *initial* report of an SIM. In other words, while there might be as long as an 8-minute period of "no" reports of SIM in a given trial of one or two minutes of signal presentation, once the subject reported a positive occurrence of stimulus-independent thought, the frequency of such reports was describable by a Poisson distribution rather than by a binomial distribution.

A procedure such as this provides some opportunity for us to see before us exactly what inherent capacities there are for processing private as well as public material, and the extent to which there may actually be inherent brain rhythms that play a role in the pattern of either sequential shifting that can occur, or in the emergence of parallel processing as well. It has also been possible to show by systematically examining content of reports in relation to whether or not the signal being presented was either visual or auditory that essentially the visual system is implicated in the production of visual SIM while the auditory system is more implicated in the production of sounds in the "mind's ear." In effect, this study lent further support to increasing evidence that privately generated phenomena do relate fairly closely to the basic imagery modalities implicated in the perceptual process as well as in the thought process (Antrobus et al., 1970).

The signal-detection model also permits the study of some degree of individual differences. Antrobus, Coleman, and Singer (1967) were able to show that subjects, already by self-report predisposed to be imaginative, were more likely as time went on to report more stimulus-independent thought than subjects who reported on a questionnaire that they were little given to daydreaming. The differences between the two groups increased over time and, indeed, so also did the number of errors. Initially, the high daydreamers reported a considerable amount of stimulus-independent thought without differing in the level of errors from the low daydreamers. As time went on, however, there was suggestion that they seemed to be preferring to respond to stimulus-independent mentation and their error rate increased significantly compared with the relatively stable rate of errors for the subjects who showed relatively little stimulus-independent mentation.

The cognitive processing model has a great many other implications that have not been examined fully. In addition to individual differences and to studies of the very process or relationship of information load from the external environment to self-generated material, we can look also at the task of processing in relation to the kind of priorities the individual may set more generally for processing life situations, whether to stress internally generated material or externally generated signals, and we can also look at the role of private material in generating specific emotional reactions. Thus, the same signal-detection task has been used in several studies to which we will refer below for establishing the implications of positive and negative affect.

In various studies directed by Antrobus and myself, we have consistently found evidence that even when persons are paid for correct signal detections, penalized for errors, forced to maintain a rapid pace of response (e.g., 1 per second), they show a fairly consistent rate of stimulus-independent thought (Antrobus et al., 1970).

An attempt was made to observe the relative frequency of two types of thought content, both unrelated to immediate task (auditory signal detections). Four persons participated in 11 consecutive, daily, 2-hour, signal-detection watches with interruptions after each 16-second trial for reports of occurrence of task-irrelevant thought. Subjects maintained an 80% accuracy detection level throughout. They reported the occurrence of stimulus-independent thought in more than 55% of the trials, a figure that was remarkably stable across the 11 daily sessions. Within the category of stimulus-independent thought, thoughts of a general nature about the experiment (but not about the detection of signals, e.g., "I'm imagining what the experimenters are doing in the next room while I'm in here") are experiment related but task irrelevant; they may be compared with more remote task-irrelevant thoughts such as "I'm picturing meeting my roommate Joe's sister next week." While experiment-related thought constituted up to 40% of all task-irrelevant thought in the first 4 sessions, it dropped off drastically during the remaining days, while more remote thought increased considerably (Antrobus et al., 1984).

In yet another study, reports of stimulus-independent thought characterized somewhat more than 50% of 80 trials by random lengths in 4 daily, signal-detection watches. Female participants reported a higher overall level of such responses; both males and females reported more task-irrelevant thought when the experimenter was of the opposite sex, but the effect was clearly greater for females (Algom & Singer, 1984).

Controlled studies of ongoing thought during signal detection watches afford a continuing rich opportunity for estimating the determinants of the thought stream. We know that the introduction of unusual or alarming information prior to entry into the detection booth (overhearing a broadcast of war news) can increase the amount of stimulus-independent thought even though accuracy of detections may not be greatly affected. A series of studies directed by Horowitz (1978) has demonstrated that specific emotional experiences of an intense nature prior to engaging in signal detections lead to emergence of material in the form of stimulus-independent ideation when thought is sampled during the detection period. Such findings have suggested a basis for understanding clinical phenomena such as "unbidden images" (Horowitz, 1978) or "pre-emptory ideation" (Klein, 1967). I believe, however, that we can go even further with such a procedure and begin to develop a systematic conceptualization of the determinants of the stream of consciousness.

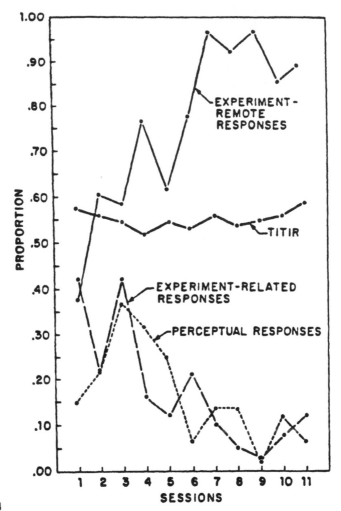

FIGURE 3

2. EXPERIMENTAL INVENTIONS AND THOUGHT SAMPLING

While the signal detection procedure gives us a powerful control over the environmental stimulus input and affords an opportunity to estimate very precisely the lengths of specific stimulus-independent thought sequences, there are somewhat less artificial methods of thought-sampling that have been increasingly employed in the development of an approach to determining the characteristics and determinants of waking conscious thought. These involve: a) asking participants to talk out loud over a period of time while in a controlled environment and then scoring the verbalization along empirically or theoretically derived categories; b) allowing the respondent to sit, recline or stand quietly for a period of time and interrupting the person periodically for reports of thought or perceptual activity; c) requiring the person to signal by means of a button press whenever a new chain of thought begins and then to report verbally in retrospect or to fill out a prepared rating form characterizing various possible features of ongoing thought.

Klinger (1977a,b, 1978, 1981) has employed thought sampling in the above forms to test a series of hypotheses about ongoing thought. He has made an interesting distinction between operant thought processes and respondent thought processes. The former category describes thoughts that have a conscious instrumental property—the solution of a specific problem, analysis of a particular issue presently confronting one, examination of the implications of a specific situation in which one finds oneself at the moment. Operant thought is active, directed, and has the characteristics of what Freud called "secondary process thinking." As Klinger has noted, it is volitional, it is checked against new information concerning its effectiveness in moving toward a solution or the consequences of a particular attempted solution, and there are continuing efforts to protect such a line of thought from drifting off target or from the intrusion of distraction either by external cues or extraneous irrelevant thought (Klinger, 1978). Operant thought seems to involve a greater sense of mental and physical effort, and it probably has the property that the neurologist Head called "vigilance" (Head, 1926); Goldstein, the "abstract attitude" (Goldstein, 1940); and Pribram and McGuinnes (1975), "effort," a human capacity especially likely to be weakened or to disappear after massive frontal brain damage. Klinger's research involving thought-sampling methods has suggested that operant thought is correlated to some degree with external situation-related circumstances. It involved higher rates of self-reports about evaluation of progress toward the goal of the thought sequence as well as of efforts to resist drift and distraction (Klinger, 1978).

Respondent thought in Klinger's terminology involves all other thought processes. These are nonvolitional in the sense of conscious direction of a sequence, and most are relatively noneffortful (Bowers, 1982). Respondent processes include seemingly unbidden images (Horowitz, 1970) or peremptory thought (Klein, 1967), which are the mental distractions one becomes aware of when trying to sustain a sequence of operant thought (analyzing the logic of a scientific or legal argument) or simply trying to concentrate on writing checks to pay bills. Most of what we consider daydreams and fantasies (and, of course, nighttime dreams) are instances of respondent thought.

The use of thought sampling in a reasonably controlled environment also permits evaluation of a variety of conditions that may influence or characterize ongoing consciousness. One can score the participants' verbalizations on dimensions such as (a) organized-sequential vs. degenerative confused thought; (b) use of imagery or related episodes or even memory material vs. logical-semantic structures; (c) reference to current concerns and unfulfilled intentions; (d) reminiscence of past events vs. orientation towards future; (e) realistic vs. improbably content, etc. A study by Pope (1978) demonstrated that longer sequences of thought with more remoteness from the participants' immediate circumstances were obtained when the respondents were reclining rather than in an interpersonal situation. Zachary (1983) evaluated the relative role of positive and negative emotional experiences just prior to a thought-sampling period. He found that intensity of experience rather than its emotional valance, and, to a lesser extent, the relative ambiguity versus clarity of the material determined recurrence in the thought stream.

Studies reviewed by Klinger, Barta and Maxeiner (1981) point to the relative importance of current concerns as determinants of the material that emerges in thought sampling. Such current concerns are defined as "the state of an organism between the time one becomes committed to pursuing a particular goal and the time one either consummates the goal or abandons its objective and disengages from the goal" (Klinger et al., 1981). Such current concerns as measured by a well-thought-out psychometric procedure make up a useful operationalization of the Freudian wish in its early (pre-libido theory) form (Holt, 1976). They may range from unfulfilled intentions to pick up a container of milk on the way home to longstanding unresolved desires to please a parent or to settle an old score with a parent or sibling. In estimating current concerns at a point in time prior to thought-sampling sessions, one obtains scale estimates of the valences of the goals, the relative importance of intentions in some value and temporal hierarchy, the person's perception of the reality of goal achievement, etc. It seems clear that only after we have explored the range and influence of such current consciously unfulfilled intentions in a sampling of the individual's thoughts, emotional and behavioral responses can we move to infer the influence of unconscious wishes or intentions.

The possibilities for controlled hypothesis-testing uses of laboratory thought sampling can be exemplified in a recent study on determinants of adolescents' ongoing thought following simulated parental confrontations (Klos & Singer, 1981). In this study, we set up a hierarchy of experimental conditions, prior to a thought sampling, which were expected to yield differential degrees of recurrence in the consciousness of the participants. We proposed that even for beginning college students, parental involvements were like to prove especially provocative of further thought. We chose to evaluate the relative role of (1) generally fufilled versus unresolved situations, the old Zeigarnick effect (Lewin, 1935); (2) a mutual non-conflictual parental interaction; (3) a confrontation or conflict with a parent that involved, however, a collaborative stance by the adult, and (4) a comparable confrontation in which the parent's attitute was clearly coercive rather than collaborative. We proposed that exposure (through a simulated interaction) to each of these conditions would yield differences in the later recurrence of simulation-relevant thoughts in the participants' consciousness. For example, we believed in general that unresolved situations would be more likely to recur than resolved ones but that, in general, the incompletion effect would be less powerful than (a) a collaborative confrontation and, especially (b) a coercive confrontation. We hypothesized that the coercive parental conflict simulation when unresolved would lead to the highest frequency of recurrence in the thoughts of the adolescents. We went a step further, however, in the light of the research just mentioned on current concerns. We proposed that a history of longstanding stress with parents would constitute a major current concern and that this factor would amplify the effect on later thought of the simulated parent interactions. Thus, frequency of recurrence in later thought of a simulated parent interaction would be highest for those participants with a history of longstanding parental conflict undergoing an unresolved coercive confrontation.

Ninety-six men and women participated in the study and were assigned (after having, some weeks earlier, reported on parental stress among other questionnaires) to one of six conditions:

a. collaborative decision-making with parent, resolved

b. collaborative decision-making with parent, unresolved

c. collaborative confrontation with parent, resolved

d. collaborative confrontation with parent, unresolved

e. coercive confrontation with parent, resolved

f. coercive confrontation with parent, unresolved

Participants engaged in carefully developed, imaginary, simulated interactions with one of their parents (seated in an "empty chair") while an experimenter read a predeveloped parental script appropriate to each situation. Three rather typical parental-child situations were used in each simulation condition. Subsequent to the simulations, subjects were taken to another room and, over a period of 20 minutes, thought samples were obtained at 20 random-interval interruptions (45-75 seconds). Their reports were tape-recorded and then scored by experimentally naive judges who rated whether verbalized content was related to definitions of the simulation settings. The participants had also made ratings of their involvement in the task, the specific emotions experienced and their relative vividness of imagery during simulation, their perception of the relative similarity of simulations to their own early experience with parents, etc. Manipulation checks failed to suggest differences other than those experimentally defined and supported the relative involvement and "reality" of the experience for this sample.

Figure 4 provides clear support for our major hypotheses. The frequency of recurrences of simulation-condition-related thoughts occur in the predicted order with the effects clearly amplified by a history of longstanding interpersonal stress with a parent. The incompletion effect is a modest one, mainly in evidence in the non-conflictual situation. It is overridden to some degree by the increasing coerciveness of the imaginary conflict situations. Of special interest is the fact that, once exposed to a simulated parent conflict, those young people who had a history of stress showed as much as 50% of their later thought reflecting this brief artificial incident. One might surmise that, if we generalize from these results, the thought world of adolescents who have had longstanding parent difficulties may be a most unpleasant domain since many conflictual chance encounters or even film or television plots might lead to recurrent thoughts to a considerable degree.

Adolescent Stress and Parental Confrontation

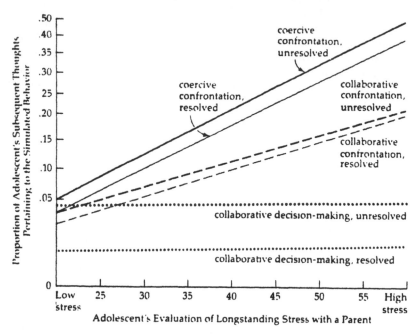

FIGURE 4 The interaction of confrontation and longstanding interpersonal stress is re-flected in the proportion of the subject's thoughts about the simulated parental confron-frontation during a twenty-minute period following the experiment. Note that unresolved confrontations produce a higher proportion of thoughts. (From Klos and Singer, 1981)

3. THOUGHT AND EXPERIENCE SAMPLING IN DAILY LIFE

It is obvious that laboratory-based methods present some difficulties because of their artificiality and also because the very controls of physical movement and re-strictions on novel sensory input which are necessary for their effectiveness may lead to overestimations of the naturally occurring fantasy and daydreaming. An approach to thought sampling that circumvents some of these problems calls for par-ticipants to carry signalling devices in pockets, purses or on pants belts as physicians do. These "beepers" go off at random during the ordinary activities of participants and they at once fill out a special card which asks for reports of activity just prior to the signal, the environmental setting, their current thoughts and emotional state. Typically these are carried for a week and they go off on the average of two-hour in-tervals permitting an accumulation of about 50-60 reports per participant. Studies by Klinger (1978), Hurlburt (1979, 1980), McDonald (1976), and a whole series direct by Csikszentmihalyi (1982; Csikszentmihalyi & Graef, 1980; Csikszentmihalyi & Kubey, 1981; Csikszentmihalyi & Larson, 1984), all demonstrate the feasibility of this method, its potential for reliable results and suitability for hypothesis testing as well as for accumulation of normative data. In a recent study with 75 adolescents in a suburban community, self-reports were obtained for 69% of the signals sent leading to an accumulation of almost 4,500 reports. Missed signals were chiefly attributable to travel outside the 50-mile signal range, "beeper" malfunctions, or sleep. Reports included such potentially censorable events as parental quarrels, sexual intimacies, or drug or alcohol abuse. Evidence for consistency and reliability are impressive in most of these studies.

Johnson and Larson (1982) used the experience sampling method with bulimics and a normative group and demonstrated that bulimics showed more dysphoric moods and also greater mood variability. They spent far more time alone at home and reported their highest levels of distress under such circumstances. One can also employ this method to evaluate the relevant influences of longstanding traits and the momentary environmental circumstances on emotional response as well as on the train of thought. In a European investigation employing a variation of this method, 24 housewives who had already taken personality tests were studied over a month. The attributions of causes of moods in various settings could be ascertained as a function of the personality characteristics of the respondent and the situation. Thus, imaginative women attributed the causes of their moods to themselves; self-confident women were more likely to attribute positive moods to their own actions rather than to others (Brandstatter, 1983). In another study, participants whose Thematic Apperception Tests pointed to greater motivation for intimacy showed more interpersonal thoughts and more positive emotional responses in interpersonal situations than did low intimacy-motive scorers based on a week-long accumulation of eight daily reports (McAdams & Constantian, 1983).

The relationship between accumulated daily reports about thought patterns and a self-report questionnaire, the Imaginal Processes Inventory (IPI) (Singer & Antrobus, 1972) was evaluated by Hurlburt (1980). He reported significant correlations between the retrospective questionnaire scales for frequent daydreaming, acceptance of daydreaming and low distractibility, and the accumulated daily reports of daydreaming based on two days of dozens of interruptions. The scale on the IPI of sexual daydreams was significantly correlated ($r=+.40$) with the accumulated record of sexual fantasies. Similarly, those persons who reported more future-oriented daydreaming on the IPI questionnaire scale actually were significantly more often likely to be engaging in such fantasies of the future ($r=+.39$) when interrupted by the electronic pager during the two days sampled.

In summary, there seems a considerable and growing availability of reasonably sophisticated measures for assessing ongoing thought in laboratory or in the field and for estimating for individuals and groups through questionnaires the trait-like patterns of current concerns, styles of daydreaming, imagery use, and absorption capacities in private experience even to the point of trance-like states. What we have not done yet is to examine in more systematic ways the links between these data derived from conscious report and the kinds of inferred unconscious schema, motivational structures and special processing patterns beneath awareness that have made up the bulk of the clinical literature on the unconscious dimension of human experience.

THE NATURE OF UNCONSCIOUS PROCESSES AND STRUCTURES

1. THE PROBLEM OF UNCONSCIOUS THOUGHT

The processes I have stressed thus far in this presentation might all be viewed as conscious in the sense that they are, under appropriate circumstances, reportable and identifiable as natural events once one engages in introspection. What, then, is unconscious? Freud, after all, on numerous occasions suggested that conscious thought was only the "tip of the iceberg." Without getting into the many subtle issues of semantics, epistemology or metaphysics, it is clear that, at the very least, our brain by an as-yet-unknown means, is capable of storing millions of bits of information, e.g., familiar words, faces, concepts, which are ultimately retrievable but of whose existence while stored we are not aware as we go about our daily business. It seems unlikely that the vast hoard of our acquired knowledge sits inertly amid some concatenation of nerve networks much as the dictionaries of computers wait in their tiny bins until activated. The seemingly sudden irruption in our dreams or waking reveries of the faces or voices of childhood or school friends or relatives suggests that storage may be a more active process. Indeed, our awareness of our stream of thought may in part be a reflection of noticing the working "machinery" of one's own brain. The problem of demonstrating the elaborate unconscious fantasies and thought content one finds in psychoanalytic inferences from patients' dream reports or symptomatic actions has proven to be a baffling one.

More recently, as social and cognitive psychologists have sought to understand how we appraise social events or organize sequences of information for later rapid retrieval and as computer scientists have tried to identify the key processes of thought in order to program artificial intellects, some important advances have occurred in our approach to unconscious process. We no longer seek to identify elaborate unconscious thought *content*, but rather to identify the structural properties of thought and the basic interactive processes that, maybe because of overlearning (as in motor acts like bicycling, swimming or just walking), operate smoothly without conscious awareness. Consider, for example, the problem of sitting down in a chair or sofa. We turn our backs and lower ourselves into the furniture while thinking of other things, carrying on elaborate conversations, watching television. For a toddler, as the ingenious films of Kurt Lewin demonstrated sixty years ago, sitting down is quite complicated because, when one turns one's back on the chair, it simply doesn't exist. One of the tragic experiences one encounters with adults suffering from Alzheimer's Disease is the difficulty they have in sitting down for apparently their once well-established schemata and image-representational structures have been disrupted, and like the toddler they are terrified about trying to sit down when facing away from a chair. These automatic processes and mental structures we take so for granted, what psychologists call our *schemata* or *scripts*, and our rules for drawing inferences or for attributing causation to correlated events which we call *heuristics* and *attributional processes* are the chief reflections of unconscious thought. Indeed, what Freud called our unconscious defense mechanisms are increasingly seen as manifestations of such more general overlearned processes of the organizing, filtering and inferential processes that we engage in continuously without awareness of their operation (Singer, 1984a).

2. SCHEMATA, SCRIPTS AND HEURISTICS

Space permits only a brief survey of some of the hypothetical constructs bearing on thought that have been identified, operationally defined, and studied in empirical research. The earlier associationist psychological theories emphasized highly specific connections between individual elements of words or events as the basis of memory. Reflecting the analyses of Bartlett, Piaget, the Gestalt School, Werner and, in clinical psychology, the personal construct theory of Kelly, modern approaches rely on somewhat more molar motions of storage structures, *schemata*. A *schema* may reflect a more or less dense (differentiated and organized) combination of an event, object or person observed in some image-like representational form often along with a verbal label or semantic categorization. Groups of schemata may gradually be further organized for efficient retrieval into lexical or functional hierarchies, e.g., dobermans into *dogs* and *animals* or into functional categories of *fierce and dangerous animals that bite* or, if one owns one, into *pets that protect* (Turk & Speers, 1983). Such organized schemata serve to point our attention towards specific objects in each new environment we confront, to help us filter the vast complexity of information we confront into managable chunks, and (as suggested by our discussion earlier of emotion) to minimize the chances that we will be startled or terrified by rapid novelty or incongruity.

Specific types of schemata may include *prototypes*, fuzzy sets of organized features that may characterize particular people or groups, e.g., men, white people, Jews, lovers, politicians, "my friends." *Self-schemata* may be the more or less differentiated descriptors and evaluative lables one stores to define a more or less bounded self-concept or a group of beliefs about self. For some people, an insult to even distant family members may be taken as an attack on one's self, while for others only slurs on one's mother or father would be so perceived.

Of special important are schemata that encapsulate elaborate action sequences, generally called *scripts*. This term originally proposed by Tomkins (1962, 1979) and employed more recently in a somewhat different form by cognitive scientists seeking to identify structures for the computer programming of normal human thought, involves the condensed representation of the actions presumably linked to thousands of typical events or setting which we confront in daily life. Thus, a computer given the sentence, "Tim had a birthday party," would usually be able to define the words of the sentence or even its grammar but would not know, as we all do, that a child's birthday party *vignette* usually involves lots of children, funny hats, noisemakers, ice cream, candles and cake, balloons, games like "pin the tail on the donkey," etc. Our personal scripts go far beyond even such a prototypic description of people, action and events to personal evaluation based on individual experiences, e.g., "My parents never made one for me" or "I hate those parties" to specific memories of events from parties one has attended which may lead to expectancies about future parties. Tomkins (1979) and Carlson (1981, 1984) have developed a theory, still be to more adequately tested, about the differential implications for information processing and behavior of *nuclear scripts* (highly charged familial or childhood schemata of action) which are either positively or negatively emotionally valuated.

It is likely that schemata about self or others, prototypes and scripts form the basis for what psychoanalysts call the transference phenomenon. The "overreactions" of intense anger, love or dependency one observes with intimates or within a psychoanalysis directed to the therapist, reflect personal scripts that are being inappropriately applied (Singer, 1974, 1984a, b.).

While the schemata represent the stored organized contents and anticipated action sequences that guide our expectancies about new events, a set of processes such as the inference and problem-solving *heuristics* proposed by Tversky and Kahnemann (1974) or a host of other biasing mechanisms involving causal attribution, assumptions of personal rather than chance control over events, favoring recent experiences as of greater importance for interpreting events, etc. (Turk & Speers, 1983). Thus, Tversky and Kahnemann in a series of ingenious studies have shown that people when confronted with a mass of new information rely on short cuts to limit search activity. These often involve *availability*, labelling or categorizing an event by the ease of retrieval of a few similar instances. If a clinician sees a patient with an "inappropriate" smiling pattern and recalls one or two recent clients who showed this feature and who were schizophrenics, the tendency to so label the new person as one may come to mind. A more careful review of one's own experiences might actually yield many negative instances where such smiling characterized normal individuals suppressing fear in public situations or even "nervous smiling" by neurotics. The *representativeness* heuristic may reflect a bias to use a cluster of traits to characterize someone, e.g., a very tall Black man one sees on the street must be a professional basketball player, without considering the base rate data or statistical odds (there are after all only about two hundred professional basketball players in the world).

Biases and thought categorization systems of this type have now been identified in literally dozens of systematic research studies. It seems increasingly clear, therefore, that human beings, along with their schemata, overlearn a variety of inferential systems that operate outside of conscious awareness to lend a tendentious quality to normal judgments and decision processes. Defense mechanisms like rationalization or denial may be special processes. Reconsiderations by cognitive-behavioral clinicians and researchers have pointed up the fact that "normal" individuals in contrast to depressives, for example, show an illusory belief in their own control over what are actually chance positive events. Such biases as well as the scripts and schema are perhaps truly unconscious processes that characterize human behavior (Meichenbaum & Gilmore, 1984).

In summary, I have tried to review a number of systematic methods and relatively new constructs that behavioral scientists have been developing to help us understand those seemingly ineffable, ongoing thought processes, conscious and unconscious, that seem so central to human experience. Space has not permitted a review of the important advances in linking such thought to specific emotions, to physiological reactivity and possibly ultimately to the immune system of the body and to the self-regulatory processes that maintain health (Jensen, 1984; Schwartz, 1982, 1983). We are, however, close to the point where, at a conference like this, one could through telemetric psychophysiological apparatus wire each of the auditors and monitor (through their signals) whenever their thoughts strayed, what emotions each experienced, whether specific content in the presentation evoked remote associations in one or a few listeners or whether, under certain circumstances, the group as a whole "tuned out" the speaker in favor of their own sequences of *task-irrelevant thought and imagery*. Indeed, one might even estimate from such a procedure by having reports toward the end whether one individual or perhaps many experienced original, stimulating or even creative new ideas as a result of the presentation. I hesitate to consider further how this review of conscious and unconscious thought might fare under such a test.

REFERENCES

Algom, D. & J. L. Singer (1984). "Interpersonal Influences on Talk-Irrelevant Thought and Imagery in a Signal-Detection Task." *Imagination, Cognition and Personality* **4**.

Antrobus, J. S. (1968). "Information Theory and Stimulus-Independent Thought." *British Journal of Psychology* **59**, 423–430.

Antrobus, J. S., R. Coleman, & J. L. Singer (1967). "Signal Detection Performance by Subjects Differing in Predisposition to Daydreaming." *Journal of Consulting Psychology* **31**, 487–491.

Antrobus, J. S., G. Fein, S. Goldstein, & J. L. Singer (1984). *Mindwandering: Time-Sharing Task-Irrelevant Thought and Imagery with Experimental Tasks.* Manuscript submitted for publication.

Antrobus J. S., & J. L. Singer (1964). "Visual Signal Detection as a Function of Sequential Task Variability of Simultaneous Speech." *Journal of Experimental Psychology* **68**, 603–610.

Antrobus, J. S., J. L. Singer, S. Goldstein, & M. Fortgang (1970). "Mindwandering and Cognitive Structure." *Transactions of the New York Academy of Sciences* (Series II), **32** (2), 242–252.

Bowers, P. B. (1982). "On *Not* Trying So Hard: Effortless Experiencing and Its Correlates." *Imagination, Cognition and Personality* **2** (1), 3–14.

Brandstatter, H. (1983). "Emotional Responses to Other Persons in Everyday Life Situations." *Journal of Personality and Social Psychology* **45**, 871–883.

Carlson, R. (1981). "Studies in Script Theory: I. Adult Analogs of a Childhood Nuclear Scene." *Journal of Personality and Social Psychology* **4**, 533–561.

Carlson, L., & R. Carlson (1984). "Affect and Psychological Manification: Derivations from Tomkins' Script Theory." *Journal of Personality* **52** (1), 36–45.

Crits-Christoph, P. (1984). *The Role of Anger in High Blood Pressure.* Unpublished doctoral dissertation, Yale University.

Csikszentmihalyi, M. (1982). "Toward a Psychology of Optimal Experience." *Review of Personality and Social Psychology*, Ed. L. Wheeler, Vol. 3. Beverly Hills, CA: Sage.

Csikszentmihalyi, M., & R. Graef (1980). "The Experience of Freedom in Daily Life." *American Journal of Community Psychology* **8**, 401–414.

Csikszentmihalyi, M., & R. Kubey (1981). "Television and the Rest of Life: A Systematic Comparison of Subjective Experience." *Public Opinion Quarterly* **45**, 317–328.

Csikszentmihalyi, M., & R. Larson (1984). *Being Adolescent.* New York: Basic Books.

Edelson, M. (1984). *Hypnosis and Evidence in Psychoanalysis.* Chicago: The University of Chicago Press.

Fein, G. G., & J. S. Antrobus (1977). "Daydreaming: A Poisson Process." *Cognitive Psychology.*

Goldstein, K. (1940). *Human Nature in the Light of Psychopathology.* Cambridge, MA: Harvard University Press.

Head, H. (1926). *Aphasia and Kindred Disorders of Speech.* Cambridge: Cambridge University Press. 2 vols.

Holt, R. R. (1976). "Drive or Wish? A Reconsideration of the Psychoanalytic Theory of Motivation." *Psychology versus Metapsychology: Psychoanalytic Essays in Memory of George S. Klein.* Eds. M. M. Gill & P. S. Holzman. *Psychological Issues,* Monograph 36. New York: International Universities Press.

Horowitz, M. J. (1970). *Image Formation and Cognition.* New York: Appleton-Century-Crofts.

Horowitz, M. J. (1978). *Image Formation and Cognition.* New York: Appleton-Century-Crofts.

Hurlburt, R. T. (1979). "Random Sampling of Cognitions and Behavior." *Journal of Research in Personality* **13**, 103–111.

Hurlburt, R. T. (1980). "Validation and Correlation of Thought Sampling with Retrospective Measures." *Cognitive Therapy and Research* **4**, 235–238.

Izard, C. E., ed. (1977). *Human Emotions.* New York: Plenum.

James. W. (1890). *The Principles of Psychology.* New York: Dover Publications, 1950. 2 vols.

Jensen, M. (1984). *Psychobiological Factors in the Prognosis and Treatment of Neoplastic Disorders.* Unpublished doctoral dissertation, Yale University.

Johnson, C., & R. Larson (1982). "Bulimia: An Analysis of Moods and Behavior." *Psychosomatic Medicine* **44**, 341–351.

Klein, G. (1967). "Peremptory Ideation: Structure and Force in Motivated Ideas." *Motives and Thought.* Ed. R. R. Hold. New York: International Universities Press.

Klinger, E. (1977a). *Meaning and Void: Inner Experience and the Incentives in People's Lives.* Minneapolis: University of Minnesota Press.

Klinger, E. (1977b). "The Nature of Fantasy and Its Clinical Uses." *Psychotherapy: Theory, Research and Practice* **14**.

Klinger, E. (1978). "Modes of Normal Conscious Flow." *The Stream of Consciousness.* Ed. by K. S. Pope & J. L. Singer. New York: Plenum.

Klinger, E. (1981). "The Central Place of Imagery in Human Functioning." *Imagery, Volume 2: Concepts, Results, and Applications.* Ed. E. Klinger. New York: Plenum.

Klinger, E., S. Barta, & J. Maxeiner (1981). "Current Concerns: Assessing Therapeutically Relevant..." *Assessment Strategies for Cognitive-Behavioral Interventions.* Ed. P. Kendall & S. Hollon. New York: Academic Press.

Klos, D. S., & J. L. Singer (1981). "Determinants of the Adolescent's Ongoing Thought Following Simulated Parental Confrontations." *Journal of Personality and Social Psychology* **41** (5), 975–987.

Kosslyn, S. M. (1981). "The Medium and the Message in Mental Imagery: A Theory." *Psychological Review* **88**, 46–66.

Langer, E. (1983). *The Psychology of Control.* Beverly Hills, CA: Sage.

Lewin, K. (1935). *A Dynamic Theory of Personality.* New York: McGraw Hill.

Luborsky, L. (1977a). "Measuring a Pervasive Psychic Structure in Psychotherapy: The Core Conflictual Relationship Theme." *Communicative Structures and Psychic Structure.* Ed. N. Freedman & S. Grand. New York: Plenum.

Luborsky, L. (1977b). "New Directions in Research on Neurotic and Psychosomatic Systems." *Current Trends in Psychology. Readings from the American Scientist.* Ed. I. L. Janis. Los Altos, CA: Kaufmann.

Mandler, G. (1975). *Mind and Emotion.* New York: John Wiley.

McAdams, D., & c. A. Constantian (1983). "Intimacy and Affiliation Motives in Daily Living: An Experience Sampling Analysis." *Journal of Personality and Social Psychology* 4, 851–861.

McClelland, D. C. (1961). *The Achieving Society.* Princeton: Van Nostrand.

McDonald, C. (1976). *Random Sampling of Cognitions: A Field Study of Daydreaming.* Unpublished master's predissertation, Yale University.

Meichenbaum, D., & J. B. Gilmore (1984). "The Nature of Unconscious Processes: A Cognitive-Behavioral Perspective." *The Unconscious Reconsidered.* Ed. K. Bowers & D. Meichenbaum. New York: Wiley.

Neville, R. C. (1981). *Reconstruction of Thinking.* Albany: State University of New York Press.

Pope, K. S. (1978). "How Gender, Solitude, and Posture Influence the Stream of Consciousness." *The Stream of Consciousness.* Eds. K. S. Pope and J. L. Singer. New York: Plenum.

Pribram, K., & D. McGuinness (1975). "Arousal, Activation and Effort in the Control of Attention." *Psychological Review* 82, 116-149.

Rychlak, J. (1977). *The Psychology of Rigorous Humanism.* New York: John Wiley.

Rychlak, J. (1981). "Logical Learning Theory: Propositions, Corollaries, and Research Evidence." *Journal of Personality and Social Psychology* 40 (4), 731–749.

Schwartz, G. E. (1982). "Cardiovascular Psychophysiology: A Systems Perspective." *Focus on Cardiovascular Psychopathology.* Eds. J. T. Cacioppo & R. E. Petty. New York: Guilford.

Schwartz, G. E. (1983). "Disregulation Theory and Disease: Applications to the Repression/Cerebral Disconnection/Cardiovascular Disorder Hypothesis." *International Review of Applied Psychology* 32, 95–118.

Segal, S. J. (1971). "Processing of the Stimulus in Imagery and Perception." *Imagery.* Ed. S. J. Segal. New York: Academic Press.

Shepard (1978). "The Mental Image." *American Psychologists* 33, 125–137.

Singer, J. L., ed. (1973). *The Child's World of Make-Believe.* New York: Academic Press.

Singer, J. L. (1974). *Imagery and Daydreaming: Methods in Psychotherapy and Behavior Modification.* New York: Academic Press.

Singer, J. L. (1984a). *The Human Personality.* San Diego, CA: Harcourt Brace Jovanovich.

Singer, J. L. (1984b). "The Private Personality." *Personality and Social Psychology Bulletin* **10**, 7–30.

Singer, J. L., & J. S. Antrobus (1972). "Daydreaming, Imaginal Processes, and Personality: A Normative Study." *The Function and Nature of Imagery.* Ed. P. Sheehan. New York: Academic Press.

Singer, J. L., S. Greenburg, & J. S. Antrobus (1971). "Looking with the Mind's Eye: Experimental Studies of Ocular Mobility during Daydreaming and Mental Arithmetic." *Transactions of the New York Academy of Sciences* **33**, 694–709.

Taylor, E. (1983). *William James on Exceptional Mental States: The 1896 Lowell Lectures.* Scribner.

Tomkins, S. S. (1962). *Affect, Imagery, Consciousness,* Vol. 1. New York: Springer.

Tomkins, S. S. (1963). *Affect, Imagery, Consciousness,* Vol. 2. New York: Springer.

Tomkins, S. S. (1979). "Script Theory: Differential Magnifications of Affects." *Nebrasksa Symposium on Motivation, 1978.* Eds. H. E. Howe, Jr. & R. A. Dienstbier. Lincoln: University of Nebraska Press.

Tucker, D. M., & P. A. Williamson (1984). "Assymetric Neural Control Systems and Human Self-Regulation." *Psychological Review* **91**, 185–215.

Turk, D. C., & M. A. Speers (1983). "Cognitive Schemata and Cognitive Processes in Cognitive-Behavioral Interventions: Going Beyond the Information Given." *Advances in Cognitive-Behavioral Research and Therapy,* Vol. 2. Ed. P. C. Kendall. New York: Academic Press.

Tversky, A., & D. Kahnemann (1974). "Judgment under Uncertainty: Heuristics and Biases." *Science* **135**, 1124–1131.

Zachary, R. (1983). *Cognitive and Affective Determinants of Ongoing Thought.* Unpublished doctoral dissertation, Yale University.

Zajonc, R. B. (1980). "Feeling and Thinking: Preferences Need No Inferences." *American Psychologist* **35**, 151–175.

MARDI J. HOROWITZ, M.D.
Professor of Psychiatry, University of California - San Francisco, San Francisco, CA 94143

Emerging Syntheses in Science:
Conscious and Unconscious Processes

Human experience is psychological and manifested through subjective knowledge and observation of behavior patterns. A given psychological phenomenon can be described not only in terms of its characteristics, but in terms of its containment within the overall flow of thought, emotion and action. An episode of unexplained panic, blushing, or a recurrent dream-like image may tend to occur in a certain state of mind, a state of mind that can be distinguished from other states of mind.

States of mind, such as episodes of queasy anxiety, can be described and classified by various theories of phenomenology. The explanation for entry into the state of mind, and for the component elements in that state of mind, is a complex one that will, in the coming years, include both neuroscience and psychological levels of explanation. The mind-body problem will in these decades be addressed in terms of transformations of information and energy, and transactive systems. Biological factors will be seen as driving psychological factors, and psychological factors as also driving biological factors.

Psychological factors, that is, causal mechanisms and transactive operations, include structures of meanings as well as processes that assess and transform information. Complex, enduring, but slowly changing structures of meaning include the self-concepts of individuals and their conceptual maps of how the self relates to others and the world. These schemata of meaning are not always available to conscious expression. The full explanation of a state of mind is difficult because such schemata of motivation and meaning operate unconsciously, and they are meaning structures rather than materially tangible ones. In addition, the involved motives are sometimes caught up in value or social conflicts. Their assessment is by clinical inference about observed patterns and the reactions to selective probes.

The neuroscience study of transitions in state has its own methodologies, and the study of conscious and unconscious reasons for changing states of mind has its methodologies. These methodologies are formidable, and scientists have tended to focus on specialization with a methodology rather than on study of a given type of human phenomenon. The study of state transition and the explanation of a state from a neuroscience level might, for example, involve use of a nuclear magnetic resonance, itself a complex cross-specialization problem ranging from physics to neurophysiology. The study of social factors and conscious mental set influences might involve specialists in perceptual process, cognitive science, and social psychology. The study of unconscious mental factors would tend to involve methodological issues of depth psychology and vital issues of how to arrive at consensual validity about second-party inferences about unconscious factors influencing a subject internally.

Progress at all levels of neural, cognitive and depth psychological sciences suggests that a convergence of explanations of a state of mind, of sets of states of mind, and of people who exhibit specific phenomena will soon be possible. In order to obtain these convergences, a revitalized focus on phenomenology is necessary, in settings that allow theoretical ranging across the mind-body problem. This work would be dissimilar to earlier philosophical contributions in that it would constantly address itself to emerging scientific methodologies in both biological and psychological arenas. This could eventually include not only the immediate explanations for a phenomenon such as a panic attack precipitated by a usually bland social stimulus but, in a new type of university setting, also a concern for historical, mythological and life-style factors that may contribute to a stiuation and how it is lived through by an individual.

J. D. COWAN
Department of Mathematics, University of Chicago, Chicago, Illinois 60637

Brain Mechanisms Underlying Visual Hallucinations

INTRODUCTION

Hallucinations are sensory images "seen" in the absence of external stimuli. They occur on falling asleep or waking up, during hypoglycemia, delirium, epilepsy, psychotic episodes, advanced syphilis, sensory deprivation, and migraine.[1] They can be triggered by photic or electrical stimulation, and by a variety of hallucinogenic drugs. It has been suggested[2] that "many important human experiences (such as dreams and visions of biblical prophets and the creative imagery of great artists are...related to hallucinations."

H. Klüver[3] made many studies of such hallucinations, especially visual ones, mainly by ingesting the drug mescaline, derived from the peyote cactus, and concluded that four types of pattern are usually observed: 1. gratings, lattices, fretworks, filigrees, honeycombs, or chessboard designs; 2. cobwebs; 3. tunnels, funnels, alleys, cones and vessels; and 4. spirals. Klüver termed these four types—*form constants*. More recent observations by R. Siegel[4] have confirmed Klüver's classification. Lattice, spiral and funnel hallucinations are shown in Figure 1.

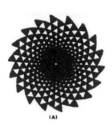

(A)   (B)   (C)

FIGURE 1   Pressure and drug-induced hallucinations. (A) Binocular pressure phos-
phene (redrawn from C. Tyler.[6]). (B), (C) LSD-induced hallucinations (redrawn from G.
Oster.[7]).

Visual hallucinations appear to be stabilized against eye movements, in much
the same way that external images are stabilized. This suggests that they are centrally, rather than peripherally located. This conclusion is supported by other observations, e.g., hallucinations induce form constants even in total darkness, and
even in blind subjects.[5]

WHAT FORM CONSTANTS LOOK LIKE IN BRAIN COORDINATES

If visual hallucinations are centrally located somewhere in the brain, it is appropriate to ask what is their geometry in terms of the coordinates of primary visual
cortex or area 17, as it is currently termed.[8] This area functions in some sense as a
cortical retina, in that there are topological maps from the eyes to area 17. However, the cortical image of a visual object is distorted. Small objects in the center
of the visual field have a much bigger representation in the cortex than do similar
objects in the peripheral visual field. The basic reason for this is that the packing
density of retinal ganglion cells falls off with increasing eccentricity in the visual
field. Since most retinal ganglion cells project, via the lateral geniculate body, to
area 17, it follows that there exists a differential representation of the visual field
in area 17. Thus, an element $dxdy$ of area 17 at the point (x, y) represents an area
$\rho r dr d\theta$ of the retinal disc at the point (r, θ), where ρ is the packing density of retinal ganglion cells. Various measurements of the packing densities of primate retinal
ganglion cells suggest that ρ may be assumed to be of the form $\rho = (\gamma^2 + \beta^2 r^2)^{-1}$
where γ and β are constants, and that $dx = \sqrt{\rho}\,dr$, $dy = \sqrt{\rho}\,r d\theta$. It follows that the
appropriate (local) area 17 coordinates are:

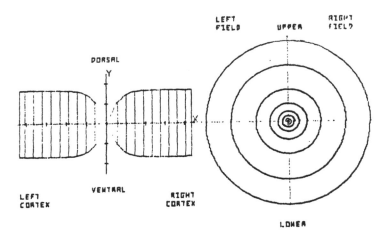

FIGURE 2 The visuo-cortical transformation. The visual field shown on the right; the corresponding cortical images of each visual hemi-field on the left. The upper right visual hemifield corresponds to the lower left cortex, and so on.

$$x = \alpha \log \left\{ \beta r + \sqrt{(1 + \beta^2 r^2)} \right\}, y = \frac{\gamma \theta r}{\sqrt{(1 + \beta^2 r^2)}}, \tag{1}$$

where (r, θ) are retinal or visual-field coordinates, (x, y) are area 17 coordinates, and α is a constant.[9]

It is easy to show that close to the center of the visual field, r small, these coordinates reduce to:

$$x = \alpha \beta r, \; y = \gamma \theta r, \tag{2}$$

polar coordinates in disguise, whereas sufficiently far away from the center:

$$x = \alpha \log \beta r, \; y = \alpha \theta. \tag{3}$$

This is the complex logarithm.[10] Its effect is to transform both dilatations and rotations of objects in the visual field into translations, parallel, respectively, to the y and x axes. Figure 2 shows the effects of the transformation (except close to the centre of the visual field). It follows that type 3 hallucinations—tunnels and funnels—become *stripe* patterns in area 17 coordinates, parallel, respectively, to the y and x axes; and that type 2 hallucinations—cobwebs—become *square lattices* parallel to the axes. Type 1 gratings and lattices still retain their lattice properties, and interestingly, class 4 hallucinations—spirals—also become stripes, the orientation of which is not parallel to either axis. Figs. 3 and 4 shows several examples of the effect of the transformation. Thus, the effect of the retino-cortical transformation is to map the form constants into either stripes of differing cortical orientation, or into lattice patterns.

HOW STRIPE AND LATTICE PATTERNS ARE FORMED

This suggests a number of interesting analogies. For example, when a fluid is heated from below, if the temperature difference between upper and lower fluid layers is sufficiently large, thermal convection occurs in the form of either hexagonal or rectangular lattices, or of stripes or "rolls" of rising and falling fluid. The hexagons are the famous Bénard convection cells,[11] and the dynamical instability which produces such patterns is known as the Rayleigh-Bénard instability. Another analogous

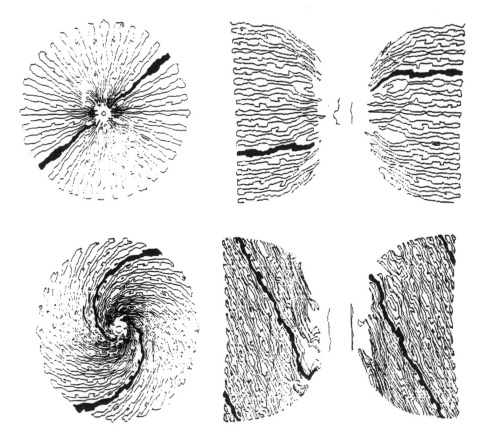

FIGURE 3 Funnel hallucination (see Fig. 1), and its cortical transformation.

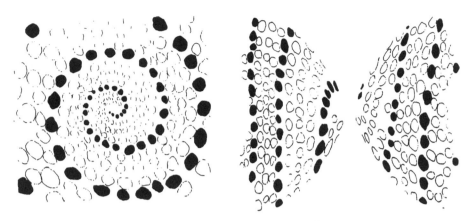

FIGURE 4   (A) Spiral hallucination (redrawn from G. Oster[7]) and its cortical transform. (B) Spiral tunnel hallucination (redrawn from Siegel[4]) and its cortical transform.

example is found in animal coat markings, e.g., in the spots and stripes of leopards and tigers. The pigmentation patterns of these species are presumed to be produced by diffusion-coupled chemical reactions which generate either lattices or stripes.[12,13] The instability which produces these pigmentation patterns was proposed first by A. Turing[14] and should perhaps be called the Turing instability. In what follows, we shall refer to the instability which produces lattice and stripe patterns, whatever their physical nature, as the Rayleigh-Bénard-Turing (RBT) instability. It is evident from our previous discussion that visual hallucinations result from the same instability, induced somewhere in the visual brain by the action of hallucinogenic drugs.

To demonstrate this, G. B. Ermentrout and I[15] analyzed the dynamics of pattern activation in model neuronal nets, and showed that lattice patterns and stripes can be generated in them by the RBT instability, in a manner completely analogous to the production of fluid convection patterns or animal coat markings. A particular net which actually generates such patterns was constructed by C. v. d. Malsburg and I.[16] It consists of sheets of neurons, each of which can *excite* both its proximal and distal neighbors and (through an interneuron) can *inhibit* its medial neighbors, as shown in Figure 5.[17] The number and strength of the contacts between neurons, and their activation thresholds, are key parameters determining the emergence of lattice and stripe patterns. They can be combining in an effective coupling parameter μ, analogous to the Rayleigh coefficient of fluid convection. It can be shown by methods outside the scope of this paper,[15,18-21] that there is a critical value μ_c, at which the resting state of the net, presumed to be zero, on the average, first becomes unstable and is spontaneously replaced by coherent patterns of large-scale activation in the form of lattices or stripes, as shown in Figure 6. The wavelength of such stripes is 2λ, where λ is the range of inhibition in the net.

It is possible to estimate the actual stripe wavelength in cortical coordinates. Consider, for example, the funnel hallucination depicted in Figure 3. Differing representations of this hallucination exist. On the average, there are about 17 stripes per hemifield (as in Figure 3). Since the cortical transform extends for some 35 mm,[22] the estimated wavelength is approximately 2 mm, whence λ, the range of cortical inhibition, is approximately 1 mm. These numbers are of considerable interest in relation to the human visual cortex: 2 mm is exactly the spacing between the blocks of cells that signal local properties of visual objects, such as position, ocularity and edge orientation, discovered by D. Hubel and T. Wiesel[6] in cats and primates (in which the spacing is, respectively, 0.3 mm and 1 mm), and termed *hypercolumns*.[23] Thus, the cortical wavelength of the stripes (and lattices) comprising hallucinatory form constants is equal to the dimensions of human hypercolumns. The analysis described above implied that inter-hypercolumnar interactions are excitatory, and intra-hypercolumnar interactions are mainly inhibitory, except for some local excitation between proximal neurons. This is consistent with what is now known about the anatomy[24] and physiology[23] of hypercolumns. Since each hypercolumnar region represents a visually distinct local patch of the visual field,[23] the circuit described above may play a fundamental role in the analysis of visual images.

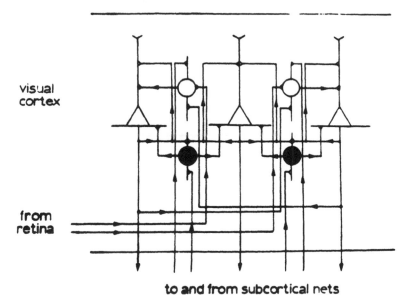

FIGURE 5 A circuit which generates stable stripe patterns. Open triangles: excitatory cells. Open circles: excitatory interneurons. Closed circles: inhibitory interneurons.

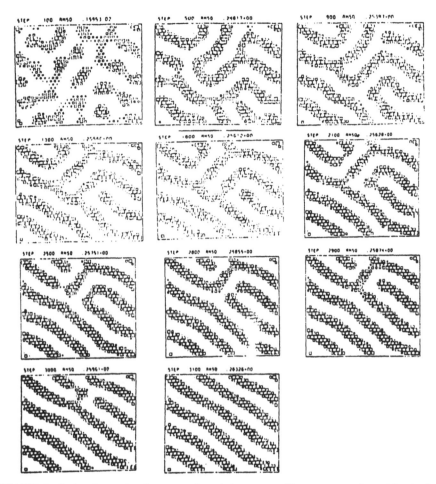

FIGURE 6 Stripe formation in a net of model neurons. The net comprises 32x32 elements arranged in a hexagonal grid. Each element excites all its proximal neighbors, inhibits its medial ones, and excites all distal cells with a strength that decreases with distance. All interactions are radially isotropic, and the boundary conditions are periodic.

PHYSIOLOGY AND PHARMACOLOGY OF CORTICAL STATES

It follows from the analysis given above, that what destabilizes the resting state of the cortex, is an increase of *excitability*. The "control" parameter μ is a measure of this excitability. I have suggested elsewhere[25] that the size of μ is determined, in part, by the actions of two brainstem nuclei, the *locus cöruleus* and the *Raphé* nucleus. The *locus cöruleus* is assumed to increase cortical excitability via the secretion of noradrenalin,[26] and the *Raphé* nucleus to decrease it via the secretion of serotonin.[27] It is known that LSD and other hallucinogens act directly on such brainstem nuclei, presumably to stimulate noradrenalin, and to inhibit serotonin, secretion. Figure 7 summarizes the details of the theory.

CONCLUDING REMARKS

The theory described above provides an account of the genesis of the simpler geometric visual hallucinations—those corresponding to the Klüver form constants—usually seen in the first stages of hallucinosis. Such form constants are shown to be generated by a cortical architecture consistent with recent anatomical and physiological discoveries.

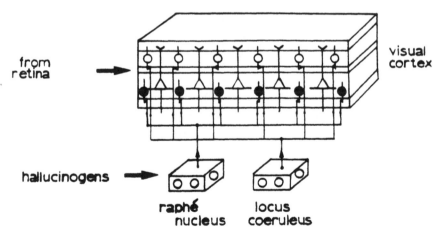

FIGURE 7   Summary of how hallucinogens are presumed to act on cortical circuits to trigger the formation of stripes and lattice patterns.

ACKNOWLEDGEMENTS

Supported in part by NATO grant 1791, and by grants from the System Development Foundation, Palo Alto, CA, and the Brain Research Foundation, University of Chicago.

REFERENCES

1. R. K. Siegel and L. J. West, eds., *Hallucinations* (New York: Wiley, 1975).
2. *Ibid*, x.
3. H. Klüver, *Mescal and Mechanisms of Hallucination* (Chicago: University of Chicago Press, 1967).
4. R. K. Siegel, *Sci. Am.* **237**, 4, 137 (1977).
5. A. E. Krill, H. J. Alpert, and A. M. Ostfield, *Arch. Opthalmol.* **69**, 180 (1963).
6. C. W. Tyler, *Vision Res.* **18**, 1633 (1978).
7. G. Oster, *Sci. Am.* **222**, 83 (1970).
8. D. H. Hubel and T. N. Wiesel, *J. Physiol. (London)* **165**, 559 (1963).
9. J. D. Cowan, *Synergetics: A Workshop*, Ed. H. Haken (New York: Springer, 1978), 228.
10. E. L. Schwartz, *Biol. Cybern.* **25**, 181 (1977).
11. S. Chandrasekhar, *Hydrodynamic and Hydromagnetic Stability* (Oxford University Press, 1961).
12. J. D. Murray, *Phil. Trans. Roy. Soc. (London) B.*, **295**, 473 (1981).
13. H. Meinhardt and A. Gierer, *J. Theor. Biol.* **85**, 429 (1980).
14. A. Turing, *Phil. Trans. Roy. Soc. (London) B.* **237**, 32 (1952).
15. G. B. Ermentrout and J. D. Cowan, *Biol. Cybern.* **34**, 137 (1979).
16. C. v. d. Malsburg and J. D. Cowan, *Biol. Cybern.* **45**, 49 (1982).
17. N. V. Swindale, *Proc. Roy. Soc. Lond. B* **208**, 243 (1980).
18. F. H. Busse, *Rep. Prog. Phys.* **41**, 1929 (1978).
19. D. H. Sattinger, *Group Theoretic Methods in Bifurcation Theory*, (New York: Springer, 1979).
20. H. Haken, *Z. Phys. B.* **21**, 105 (1975).
21. M. Golubitsky, *Colloque Int. Probs. Math. Phénom. Nat.* (Holland: Redidel, 1983).
22. R. Hartwell and J. D. Cowan (unpublished).
23. D. H. Hubel and T. N. Wiesel, *J. Comp. Neurol.* **158**, 295 (1974).
24. J. Szentagothai, *Neurosci. Res. Bull.* **12**, 3 (1974).
25. J. D. Cowan, *Int. J. Quant. Chem.* **22**, 1059 (1982).
26. N.-S. Chu and F. Bloom, *J. Neurobiol* **5**, 527 (1974).
27. L. L. Iverson, *Sci. Am.* **241**, 3, 134 (1979).

ALWYN C. SCOTT
Center for Nonlinear Studies, Los Alamos National Laboratory, Los Alamos, NM 87545

Solitons in Biological Molecules[1]

WHAT IS A SOLITON?

At the present time, it is generally accepted that the soliton concept plays a significant role in understanding the dynamical behavior of localized or self-trapped states
in condensed matter physics, plasma physics and hydrodynamics. To my knowledge,
the first example of such a self-trapped state in condensed matter physics was the
"polaron" suggested in 1933.[1] In this case an electron moves through a crystal as
localized wave function rather than an extended Bloch state. Since the electron is
localized, it polarizes the crystal in its vicinity, thereby lowering its energy, which
keeps it localized.

In these comments the term *soliton* is used in a generic sense to denote all
examples of dynamic self-trapping; thus, a polaron is a soliton, but a soliton is not
necessarily a polaron.

[1] February 1984 (April 1984, Revised)

To understand the concept of self-trapping, it is helpful to consider some simple wave equations. Support $u(x, t)$ is some variable (amplitude of a water wave, for example) which depends upon distance (x) and time (t). A linear wave equation expressing this dependency might be

$$\frac{\partial u}{\partial t} + \frac{\partial^3 u}{\partial x^3} = 0. \tag{1}$$

For typographical convenience I will use a subscript notation for patrial derivatives; thus (1) becomes

$$u_t + u_{xxx} = 0. \tag{1'}$$

An elementary solution of this linear wave equation can be written as the complex sinusoid

$$u_e(x, t) = exp(ikx + i\omega t) \tag{2a}$$
$$= cos(kx + \omega t) + i\,sin(kx + \omega t) \tag{2b}$$

for which the frequency ($\omega = 2\pi$/temporal period) and the wave number ($k = 2\pi$/spatial period) are related by the requirement that

$$\omega = k^3. \tag{3}$$

The elementary solution in (2) is nonlocalized; it extends with equal amplitude from $x = -\infty$ to $x = +\infty$. To represent a pulse-like or localized solution, we can take advantage of the fact that (1) is *linear* so a sum of elementary solutions with different amplitudes is also a solution. A general way to write such a sum is as the Fourier integral

$$u(x, t) = \int_{-\infty}^{\infty} F(k)\,exp(ikx + i\omega t)dk \tag{4}$$

for which any initial condition $u(x, 0)$ can be matched by choosing $F(k)$ to satisfy

$$u(x, 0) = \int_{-\infty}^{\infty} F(k)\,exp(ikx)dk. \tag{5}$$

A point of constant phase on each of the elementary (Fourier) components of (4) travels with the phase velocity

$$v_{ph} = \frac{\omega}{k} = k^2. \tag{6}$$

Since this velocity is different for each component, the initial pulse shape, $u(x, 0)$, will spread out or *disperse* as time increases.

Now suppose that (1) is augmented to become the *nonlinear wave equation*

$$u_t - uu_x + uxxx = 0. \tag{7}$$

An elementary solution of this equation is

$$u_e(x, t) = 3v \, sech^2 \left\{ \sqrt{v}(x - vt)/2 \right\} \tag{8}$$

for which the propagation velocity, v, can be real number ≥ 0. The elementary solution in (8) is *localized*. A pulse with initial shape

$$u(x, 0) = 3v \, sec^2(\sqrt{v}x) \tag{9}$$

does not disperse with time but evolves with the stable pulse shape given by (8). It is as if the nonlinear term (uu_x) in (7) acts to counter effects of the dispersive term (u_{xxx}).

The term *soliton* was coined in 1965 to denote the pulse-like solution of (7) that is displayed in (8).[2] Since that time, applied scientists in many areas of research (e.g., hydrodynamics, optics, plasma physics, solid state physics, elementary particle theory and biochemistry) have begun to consider nonlinear features of their respective wave problems and to take seriously the pulse-like elementary solutions that emerge. These localized entities serve as carriers of mass, electric charge, vibrational energy, electromagnetic energy, magnetic flux, etc. depending on the particular context, and the term soliton is now used to indicate any and all of them. A number of books on solitons are listed at the end of these comments for the reader who wishes to learn more about this growing area of research.

It is important to emphasize that the soliton concept is fundamentally *nonequilibrium* in nature. The energy that is localized in the stable, pulse-like solution given by (8) is prevented by the nonlinearity in (7) from redistributing itself into a small amplitude and nonlocalized solution of (1).

DAVYDOV'S SOLITON

A fundamental problem in biochemistry is to understand how metabolic energy is stored and transported in biological molecules.[3] An interesting candidate is the amide-I (or CO stretching) vibration in protein. This vibration has a quantum energy of about 0.2 ev which is appropriate to store or transport the 10 kcal/mole (.43 ev) of free energy released in the hydrolysis of ATP. However, a linear theory won't fly. If amide-I vibrational energy is assumed to be localized on one or a few neighboring peptide groups at some time, it will rapidly disperse and distribute itself uniformly over the molecule. The cause of this dispersion is the dipole-dipole interactions between vibrating and nonvibrating peptide groups. This interaction (which is similar to the interaction between the transmitting and receiving antennae of a radio system) requires that initially localized energy become nonlocalized in about a picosecond, a time that is much too short for biological significance.

To see this effect in more detail, consider the alpha-helix shown in Fig. 1. An important variable is the probability amplitude $a_{n\alpha}$ for finding an amide-I vibrational quantum in the peptide group specified by the subscripts n and α. Thus, the probability of a vibrational quantum being at the (n, α) peptide group if $\mid a_{n\alpha} \mid^2$ where n is an index that counts turns of the helix and $\alpha (= 1, 2 \, or \, 3)$ specifies one of the three peptide groups in each turn.

If dipole-dipole interactions were not present, $a_{n\alpha}$ would obey Schrödinger's time dependent equation

$$i \, \hbar a_{n\alpha,t} = E_o a_{n\alpha} \tag{10}$$

where E_o is the energy of an amide-I quantum. Equation (10) has the solution

$$a_{n\alpha} \propto exp(-iE_o t/ \hbar). \tag{11}$$

Thus $\mid a_{n\alpha} \mid^2$ would remain constant over time and any initial localization of amide-I vibrational energy would not change.

Taking account of dipole-dipole interactions, (10) becomes

$$i \, \hbar a_{n\alpha,t} = E_o a_{n\alpha} - J \left(a_{n+1,\alpha} + a_{n-1,\alpha} \right) + L \left(a_{n,\alpha+1} + a_{n,\alpha-1} \right) \tag{12}$$

where J is the strength of the longitudinal, nearest neighbor, dipole-dipole interaction and L plays a corresponding role for lateral interaction. Equation (12) is highly dispersive. Initially localized vibrational energy would quickly spread longitudinally through J and laterally through L.

In 1973 Davydov and Kislukha proposed a nonlinear mechanism that might prevent energy dispersion in (12).[4] This mechanism involves interaction with longitudinal sound waves or stretching of the hydrogen bonds (see Fig. 1). Without this nonlinear effect, longitudinal sound waves would be governed by the equation

$$M z_{n\alpha,tt} - K \left(z_{n+1,\alpha} - 2z_{n\alpha} + z_{n-1,\alpha} \right) = 0 \tag{13}$$

where $z_{n\alpha}$ is the longitudinal displacement of the (n, α) peptide group, M is the mass of a peptide group plus residue, and K is the spring constant of a hydrogen bond.

The specific nonlinear effect considered by Davydov was the effect of stretching a hydrogen bond on the amide-I quantum energy (E_o). If R is the length of the amide's hydrogen bond, this effect can be expressed as the nonlinear parameter

$$\chi = \frac{dE_o}{dR}. \tag{14}$$

Since E_o can be measured in joules and R in meters, χ has the units of newtons. Its value has been calculated (using self-consistent field methods) as[5]

$$\chi = 3 - 5 \times 10^{-11} \text{ newtons.}$$

It has also been determined experimentally (from a comparison of hydrogen-bonded polypeptide crystals with different bond lengths and amide-I energies) as[6]

$$\chi = 6.2 \times 10^{-11} \text{ newtons.}$$

Augmenting (12) and (13) by the interaction expressed in (14), led Daydov to the nonlinear wave system

$$
\begin{aligned}
i\,\hbar a_{n\alpha,t} &= \{E_o + \chi(z_{n+1,\alpha} - z_{n\alpha})\}\, a_{n\alpha} - \\
&\quad J(a_{n+a,\alpha} + a_{n-1,\alpha}) + L(a_{n,\alpha+1} + a_{n,\alpha-1}) \tag{15a}
\end{aligned}
$$

$$M z_{n\alpha,tt} - K(z_{n+1,\alpha} - 2z_{n\alpha} + z_{n-1,\alpha}) = -\chi \left(|a_{n\alpha}|^2 - |a_{n-1,\alpha}|^2 \right). \tag{15b}$$

As previously noted, the subscripts n and α specify a particular peptide group; thus n counts turns of the helix and $\alpha(= 1, 2 \text{ or } 3)$ denotes one of the three peptide groups in each turn (see Fig. 1). The subscripts t and tt denote first and second derivatives with respect to time. Note that the only changes from (12) and (13) are to take account of the force due to stretching of the hydrogen bond in the $\chi(z_{n+1,\alpha} - z_{n\alpha}a_{n\alpha}$ term of (15a) and the corresponding source term $-\chi \left(|a_{n\alpha}|^2 - |a_{n-1,\alpha}|^2 \right)$ in (15b). It is important to observe that *each parameter* ($\hbar, E_o, \chi, J, L, M$ and K) which appears in (15) has been independently determined. Thus, a study of the dynamical behavior permits no parameter adjustment whatsoever, a rather unusual situation in biological science.

A detailed analytical study of (15) has shown that solitons do indeed form.[7] If amide-I vibrational energy is localized on one or a few neighboring peptide groups, then the right-hand side of (15) is nonzero and acts as a source of longitudinal sound. This longitudinal sound, once created, reacts, through the term $\chi(z_{n+1,\alpha} - z_{n\alpha})$, in (15a) as a potential well to trap the localized amide-I vibrational energy and prevent its dispersion by the effects of dipole-dipole interactions.

Davydov's soliton concept is rather similar to the polaron.[1] For the polaron, localized electronic charge distorts the lattice in its vicinity, lowering its energy and thereby trapping it. For Daydov's soliton, localized amide-I vibrational energy distorts the lattice in its vicinity, lowering its energy and thereby trapping it.

An extended analytical and numerical study of (15) has also shown that a *threshold* for soliton formation must be considered.[8] If a certain amount of amide-I

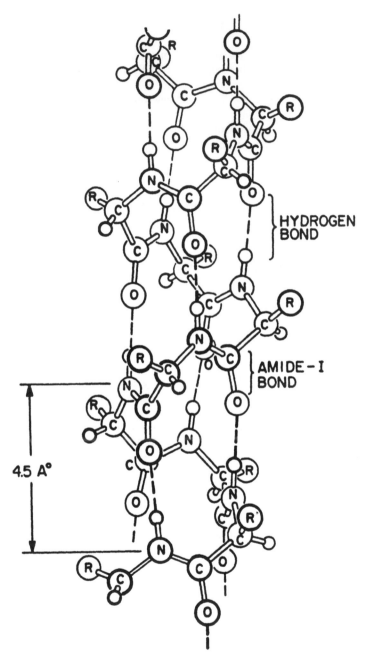

FIGURE 1   The alpha-helix structure in protein.

vibrational energy is initially placed at one end of an alpha-helix, the nonlinear parameter χ may or may not be large enough to hold a soliton together. In other words, a realistic quantity of amide-I vibrational energy, acting through χ on the right-hand side of (15b), may (or may not) create enough longitudinal sound to react, again through χ in (15a), to a degree sufficient to support a soliton. This is a key scientific issue which can be expressed as the

QUESTION: Is the experimentally determined value of χ sufficiently large to hold the amount of energy released in ATP hydrolysis together as a soliton?

To answer this question, I have conducted a numerical study of a system of equations similar to (15) but including ten additional dipole-dipole interaction terms in order to avoid underestimating the effects of dispersion.[8] A typical result of this study is shown in Figure 2 for which the computational parameters are as follows:

1. The length of the helix was chosen to be 200 turns (which corresponds to the length of an alpha-helix fiber in myosin).
2. The total amide-I vibrational energy was assumed to be two quanta (ca. 0.4 ev) or about the free energy released in ATP hydrolysis.
3. At the initial time these quanta were placed in the first turn of the helix.
4. The calculations were continued for a time of 36 ps.
5. Increasing values of the nonlinear parameter were used for each dynamical calculation.

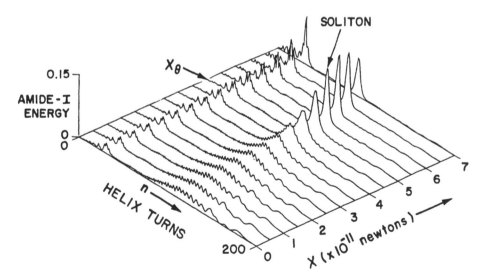

FIGURE 2   Solutions of equations similar to (15) that demonstrate formation of a Davydov soliton for the nonlinear parameter $\chi > 4.5 \times 10^{-11}$ newtons at $t = 36$ picoseconds. See text for additional details.

From Fig. 2 we see that with $\chi = 0$ the initially localized amide-I vibrational energy becomes, after 36 ps, dispersed over the entire molecule. This is to be expected since, with $\chi = 0$, one is essentially integrating (12). As χ is increased from zero, there is little change in the dynamical behavior until $\chi \cong 4.5 \times 10^{-11}$ newtons. Above this threshold value, a soliton forms with the "squared hyperbolic secant" shape that we met previously in (8).

Since the spirit of the calculation was to avoid underestimating dispersive effects, one must conclude that the threshold level for χ to achieve soliton formation is

$$\chi_\theta < 4.5 \times 10^{-11} \text{newtons.}$$

This value compares very favorably with the above-noted experimental and calculated values for χ.

Thus the answer to the *QUESTION* is "Yes."

It is important to emphasize the difference between calculations based on (15) and conventional molecular dynamics calculations.[9] Conventional calculations explore molecular dynamics in the vicinity of thermal equilibrium. At room temperature this implies

$$| a_{n\alpha} |^2 \sim exp[-E_{sol}/kT]$$
$$\sim 10^{-4} \tag{16}$$

so (15a) can be ignored and the right-hand side of (15b) set to zero. Near thermal equilibrium, therefore, (15) reduces to (13) and a conventional molecular dynamics calculation would describe only the propagation of longitudinal sound waves.

But, as has been emphasized above, the soliton is a *nonequilibrium* concept. Values of $| a_{n\alpha} |^2$ displayed in Fig. 2 are much larger than is indicated in (16). The point of Davydov's theory[4,7] is that amide-I vibrational energy can remain organized (self-trapped) in these soliton states for times long enough to be of biological interest.

What about losses of amide-I vibrational energy to water? This is an important question because the HOH bending vibration of water absorbs strongly near the amide-I frequency. Some perspective on the answer to this question can be obtained by considering soliton propagation on an alpha-helix that is completely immersed in water.[10] Symmetric solitons (for which $a_{n1} = a_{n2} = a_{n3}$) would have a lifetime of about 500 ps. Antisymmetric solitons (for which $a_{n1} + a_{n2} + a_{n3} = 0$) have a much smaller net dipole moment and are expected to have a lifetime much longer than 500 ps.

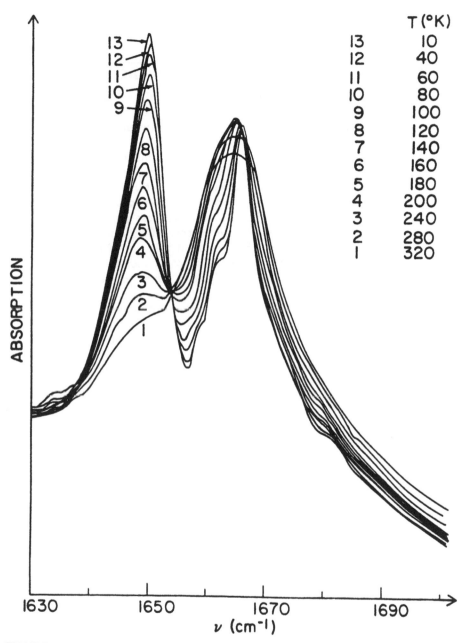

FIGURE 3 Infrared absorption spectra of polycrystalline acetanilide. The peak at 1650 cm^{-1} is interpreted as absorption by a self-trapped (soliton) state similar to that described by Davydov. The intensity of this peak depends on temperature as $[1 - exp(-\hbar\omega/kT)]^2$ as discussed in the text. (Data by E. Gratton.[6])

DAVYDOV-LIKE SOLITONS IN CRYSTALLINE ACETANILIDE

We have seen that the physical parameters of alpha-helix are such that one can expect soliton formation at the level of energy released by ATO hydrolosis. It is now appropriate to ask whether there is any direct experimental evidence for such self-trapped states.

An interesting material to consider is crystalline acetanilide ($CH_3CONHC_6H_5)_x$ or ACN. This is an organic solid in which chains of hydrogen-bonded peptide groups run through the crystal in a manner quite similar to the three chains of hydrogen-bonded peptide groups seen in Fig. 1. Around 1970 Careri noted that the peptide bond angles and lengths in ACN are almost identical to those in natural protein, and he began a systematic spectral study to see whether ACN displayed any unusual properties that might shed light on the dynamical behavior of natural proteins. He soon found an "unconventional" amide-I absorption line at 1650 cm^{-1} that is red shifted from the conventional peak by about 15 wavenumbers.[11] This effect is displayed in Fig. 3.

Since the factor group of ACN is D_{2h}^{15}, the three IR active modes have symmetries B_{1u}, B_{2u} and B_{3u} which correspond to the requirement that the x, y and z components, respectively, of the amide-I transition dipole moments in phase. The splitting of these modes from dipole-dipole interactions is less than 3 wavenumbers[12]; thus the three components (1665, 1662 and 1659 cm^{-1}) of the high-frequency band in Fig. 3 (seen most clearly at 10K) are identified as the B_{2u}, B_{1u} and B_{3u} modes, respectively.[6] This assignment leaves no place for the 1650 cm^{-1} band, yet N^{15} substitution experiments, polarized Raman measurements and polarized IR absorption measurements clearly identify it as amide-I. Careful measurements of x-ray structure and of specific heat as functions of temperature rule out a first-order phase transition and studies of the Raman scattering below 200 cm^{-1} as a function of temperature preclude a second-order phase transition. Other considerations make an explanation of the 1650 cm^{-1} band based on Fermi resonance or on localized traps unlikely.[6] Thus, the assignment of the 1650 cm^{-1} band to a self-trapped (soliton) state is quite attractive.

In such an assignment the self-trapping is assumed to arise through the interaction of localized amide-I vibrational energy with low frequency phonons. This interaction displaces the ground states of the low frequency phonons slightly leading to a Franck-Condon factor[2]

$$F \geq exp - \frac{\Delta E}{4 \; \hbar\omega} \tag{17}$$

---

[2] In calculating the transition rate for optical absorption by a soliton, one must include the square of the inner product of the phonon wave functions before and after the transition (Fermi's "golden rule"). This "Franck-Condon factor" can be small (forbidding the transition) if the response frequency of the phonon field (ω) is sufficiently small.

where ΔE is the binding energy of the soliton (15 cm^{-1}) and $\hbar\omega$ is the energy of the low frequency phonon.[6] For an *acoustic phonon*, as in (15), $\hbar\omega \ll \Delta E$ and $F \cong 0$. Thus, direct optical absorption is forbidden for solitons that are self-trapped through interaction with acoustic phonons.[7]

At non-zero temperatures, the probability amplitude for the ground state of a low frequency phonon is proportional to $[1 - exp(- \hbar\omega/kT)]^{1/2}$. Since the Franck-Condon factor involves the *square* of the *product* of the ground states before and after the transition, it is proportional to $[1 - exp(- \hbar\omega/kT)]^2$. Fitting this function to the intensity dependence of the 1650 cm^{-1} peak in Fig. 3 implies $\hbar\omega = 130$ cm^{-1} which is close to several optical modes of ACN.[13] Thus, the 1650 cm^{-1} peak can be interpreted as self-trapping through interaction with optical phonons for which the low-temperature Franck-Condon factor $F \cong 1$.

MORE JARGON

In the foregoing discussion I have repeatedly emphasized that the soliton is a nonequilibrium object. Eventually one expects the effects of dissipation to bring its amplitude back to the thermal level suggested by (16). But that is only part of the story.

Another type of soliton appears upon examination of the nonlinear wave equation

$$u_{xx} - u_{tt} = sin\,u \tag{18}$$

which was suggested in 1939 as a model for the propagation of a dislocation in a crystal.[14] This equation has the localized solutions

$$u(x, t) = 4\,tan^{-1}\left[exp\left(\pm\frac{x - vt}{\sqrt{1 - v^2}}\right)\right]. \tag{19}$$

With the "+" sign

$$u \to 2\pi \text{ as } x \to +\infty \text{ and } u \to 0 \text{ as } x \to -\infty \tag{20a}$$

while with the "-" sign

$$u \to 2\pi \text{ as } x \to -\infty \text{ and } u \to 0 \text{ as } x \to +\infty. \tag{20b}$$

Any velocity (v) of magnitude less than unity can be used in (19) and the effect of dissipation is to force this velocity to zero. When this happens, however, the solution does not disappear: it continues to satisfy the boundary conditions (20). Since such boundary conditions are called "topological constraints" by mathematicians, (19) is termed a *topological soliton*. It is natural to describe moving or stationary phase boundaries (Bloch walls in a ferromagnet, for example) as topological solitons. The fact that $u(x, t)$ itself cannot be normalized is not a problem in such applications because energy density depends only upon derivatives of u with respect to x and t.

By default, the solitons that we have previously considered as solutions of (8) and (15) are called *nontopological solitons*. Under the action of dissipation, the amplitude of a nontopological soliton will decay to the thermal equilibrium value.

Topological solitons can, however, be created and destroyed in pairs. To show this, one can construct a solution of (18) as

$$u(x, t) = 4 \tan^{-1} \left[exp \left(+\frac{x + x_o - vt}{\sqrt{1 - v^2}} \right) \right] + 4 \tan^{-1} \left[exp \left(-\frac{x - x_o - vt}{\sqrt{1 - v^2}} \right) \right] \quad (21)$$

where $x_o \gg 1$. The first term on the right-hand side of (21) is a topological soliton moving with velocity v in the x-direction and, at $t = 0$, located at $x = -x_o$. the second term is an "anti-topological soliton" moving with velocity v in the x-direction and, at $t = 0$, located at $x = +x_o$. Now the topological constraint is

$$u \to \quad \text{as } x \to \pm\infty \quad (22)$$

which is the same as that for the vacuum solution $u(x, t) = 0$. Thus, one can imagine a continuous deformation of the vacuum solution into (21) that does not disturb the boundary conditions.

Since the terms "topological soliton" and "anti-topological soliton" are somewhat inconvenient, soliton buffs often use the terms *kink* and *antikink* instead.

Considering $u(x, t)$ to be an angle, it is clear from the first term of the right-hand side of (21) that a kink carries a twist of $+2\pi$ which is observed as one goes from a large negative value of x to a large positive value. Likewise from the second term on the right-hand side of (21), an antikink is seen to carry a twist of -2π. A solution composed of M kinks and N antikinks is seen to carry a twist of $(M-N)2\pi$. This net twist is equal to $u(+\infty, t) - u(-\infty, t)$ which is a constant of the motion.

Often one speaks of "topological charge" which is equal to the twist measured in units of 2π. Thus, a kink has a topological charge of $+1$, an antikink has a topological charge of -1, and the net topological charge is a constant of the motion. It follows that kinks and antikinks must be created and destroyed in pairs in order to keep the net topological charge constant. This is analogous to the creation or destruction of positrons and electrons in pairs (out of or into the vacuum) without changing the net electric charge.

SOLITONS IN DNA

As a long biopolymer with complex dynamic behavior which is poorly understood, DNA offers attractive possibilities for nonlinear pulse propagation. In seeking such behavior, it is prudent to begin with an understanding of the underlying linear wave behavior. Early in these comments, we became familiar with the linear wave Eq. (1) before considering the effects of nonlinear augmentation to (7). Likewise in thinking about Daydov's soliton, we introduced the linear equation describing dispersion of amide-I vibrational energy (12) and the linear equation for lingitudinal sound waves (13) before considering their augmentation to the nonlinear coupled system (15). Thus, the place to start a study of nonlinear wave dynamics on DNA is with a firm understanding of the linear behavior. Some progress has recently been made toward understanding the propagation of linear acoustic waves on DNA based upon Brillouin scattering measurements of wave speeds.[15] This work indicates rather high longitudinal wave speeds: $\sim$ 3800 meters/second in dry fiber (A-conformation) and $\sim$ 1800 meters/second in wet fiber (B-conformation). Extensive normal mode calculations for both acoustic and optic modes show that the acoustic wave speeds noted above require long-range forces in the A-conformation and probably also in the B-conformation.[16]

The first specific suggestion of soliton states on DNA invoked (18) to describe the thermal breaking of hydrogen bonds between base pairs which, in turn, was supposed to explain experimental measurements of hydrogen deuterium exchange rates.[17] This idea has been developed in some detail as the "dynamic plane base-rotator model" for which chaotic behavior has been investigated.[18] Linear calculations of hydrogen bond stretching modes near the end of a strand,[19] seem to have found experimental confirmation.[20]

From both theoretical[21] and numerical studies,[22] it is clear that the thermal excitation of kinks and antikinks plays a key role in phase transitions. Starting with this insight, Krumhansl and Alexander are constructing a dynamical model of DNA with topological solitons for which the state of the system approaches (say) the A-conformation as $x \rightarrow +\infty$ and the B-conformation as $x \rightarrow -\infty$.[23] Rather than attempting to follow the motions of all atoms, they are isolating a few significant coordinates for dynamic simulation. Their present effort is directed toward making this model consistent with the linear description.[16]

Sobell has proposed for DNA the nontopological structure shown in Fig. 4.[24] Scanning from left to right, this figure shows a kink transition from B-conformation into a central region of modulated β alternation in sugar puckering along the polymer backbone, followed by an antikink transition back to the B-conformation. The β premelted core exhibits a breathing motion facilitates drug intercalation and it may provide a nucleation center for RNA polymerase-promoter recognition. This is a chemical model that is suggested by the physical equations, but a direct connection has yet to be shown.

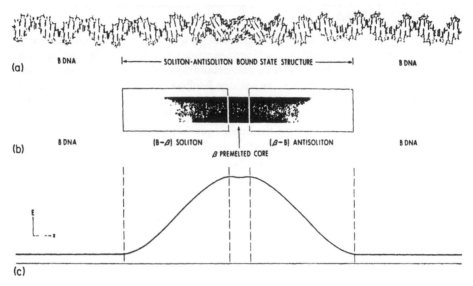

FIGURE 4   Sobell's suggestion for a kink-antikink soliton structure in B-conformation DNA. (a) Molecular structure in the vicinity of the soliton. (b) schematic representation of three regions of the soliton. (c) Energy density (E) as a function of position (x).

Some appreciation for the complex dynamic behavior to be expected in DNA is obtained from recent molecular dynamics simulations.[25] These followed the motions of the 754 atoms in the dodecamer $(CGCGAATTCGCG)_2$ and the motions of the 1530 atoms in the 24-bp fragment $(A)_{24}(T)_{24}$ for times up to 96 ps. (48,000 time steps). Motions that could encourage drug intercalation were observed. Ultimately a confirmation of the soliton structure suggested in Fig. 4 must be based upon such simulations.

THE POLARON

Let us briefly return to the self-trapping of electric charge which was mentioned at the beginning of these comments. Landau's original suggestion[1] was discussed in detail by Pekar[26] (who seems to have coined the term "polaron" for the localized electron plus lattic distortion), by Fröhlich,[27] and by Holstein.[28] Since 1970 the polaron has been widely studied in condensed matter physics.

Charge transport is often an important function in biological molecules, and, as Davydov has noted,[7] the polaron may be involved. Analytically the description of charge transport on an alpha-helix is identical to (15); one merely interprets $\underline{a}$ as the electronic wave function and $\underline{z}$ as a longitudinal polarization. Recently, a theory for proton transport in purple membrane that is based upon a topological soliton (kink) similar to (19) has been developed.[29]

SOLITONS OF ELECTRICAL POLARIZATION

Fröhlich has also suggested that polarization effects might play an important role in determining the conformational states of biological molecules.[30] He considered, as a simple example, a sphere which could be elastically deformed into an ellipsoid of eccentricity η. The elastic energy of deformation would be $1/2\, a\eta^2$ where $a > 0$. The self-energy of polarization would be $1/2\, b(1 - c\eta)P^2$ where P is the polarization vector and c may be positive. Thus, the sum of these energies

$$U = \frac{1}{2}a\eta^2 + \frac{1}{2}b(1 - c\eta)P^2 \tag{23}$$

is a minimum at the eccentricity

$$\eta = \frac{1}{2}\left(\frac{cb}{a}\right)P^2. \tag{24}$$

Near this minimum

$$U = \frac{1}{2}b\,P^2\left(1 - \frac{1}{4}\left(\frac{bc^2}{a}\right)P^2\right) \tag{25}$$

Minimizing again with respect to P implies that the original sphere might deform itself into an ellipsoid with polarization $P = \frac{2}{c}\sqrt{a/b}$ and eccentricity $\eta = 2/c$. Such a conformational change could be induced, for example, in intrinsic membrane proteins through the action of the transmembrane potential.[31]

The above discussion describes an "electret" which is the electrical analog of a magnet. Bilz, Büttner and Fröhlich have noted that 90% of the materials that display this property are oxides, and oxygen is a common constituent of organic materials.[32] They have proposed nonlinear wave equations involving interactions of polarization waves with phnons and leading to solitons of both the kink and the antikink varieties. The biological significance of these solitons has been briefly discussed.

Recently, Takeno has reformulated the quantum mechanical basis for Davydov's soliton theory.[33] His results reduce to (15) when $J \ll E_o$, but his dynamic equations describe polarization of the alpha-helix rather than the probability amplitude for finding a quantum of amide-I vibrational energy. Thus, this picture nicely complements the ideas developed by Fröhlich.

CONFORMONS

In 1972 the term conformon was coined independently by Volkenstein[34] and by Green and Ji[35] to describe a common mechanism for both enzymic catalysis and biological energy coupling. Volkenstein's conformon was defined as a nonlinear state composed of an electron plus local deformation of a macromolecule; thus it is quite similar to a polaron. The conformon of Green and Ji was defined as "the free energy associated with a localized conformational strain in biological macromolecules" and characterized as follows.

"i) The conformon is mobile. The migration of the conformon requires a relatively rigid protein framework such as the α-helical structure.

"ii) The conformon differs from the generalized electromechanochemical free energy of protein conformational strains in the sense that the conformon has the property of a packet of energy associated with conformation strain localized within a relatively small volume compared with the size of the supermolecule.

"iii) The path of the conformon migration need not be rectilinear but will be dependent on the 3-dimensional arrangement of the linkage system.

"iv) The properties of the conformon are believed to be intimately tied in with the vibrational coupling between adjacent bonds in polypeptide chains."

From this characterization, the conformon of Green and Ji is seen to be rather closely related to the Davydov soliton. This relationship has recently been discussed in a paper by Ji[36] which emphasized the ability of the conformon theory to explain both membrane-associated and membrane-independent coupled processes whereas the chemiosmotic theory[37] requires a coupling membrane to generate a transmembrane electrochemical gradient of protons. A general quantum mechanical formalism for both the Volkenstein and the Green-Ji conformons has also been developed.[38]

FRÖHLICH'S THEORY

In 1968 Fröhlich introduced a biophysical concept that has stimulated a number of experimental investigations. He assumed a collection of z oscillatory modes with frequencies $\omega_1 < \omega_2 < \cdots < \omega_j < \cdots < \omega_z$ which do not interact directly with each other but can exchange quanta with a heat bath. Nonlinear interactions between the modes arise from simultaneous absorption and emission of quanta with different energies. Metabolic energy input to the system was represented by supposing each mode to receive input power represented by the parameter s. Under these conditions, the steady-state number of quanta in each mode is given by[39]

$$n_j = \frac{A}{exp\left[\left(\hbar\omega_j - \mu\right)/kT\right] - 1} \tag{26}$$

where $A > 1$ and $\mu > 0$, but as $s \to 0$, $A \to 1$ and $\mu \to 0$.

Thus, for $s = 0$ (no metabolic pumping of the modes), (26) reduces to the expected Bose-Einstein expression for the thermal equilibrium number of quanta in an harmonic oscillator. As s is increased from zero,

$$\mu \to \hbar\omega_1, \tag{27}$$

and (26) implies that the number of quanta in the mode of lowest frequency becomes very large. This effect is similar to Bose-Einstein condensation in superfluids and in superconductors except that the order arises when the metabolic drive (s) is made sufficiently large rather than by lowering the temperature.

This is a generic idea, but the details are important. As a rough estimate of the set of frequencies $\{w_j\}$ that might be involved in a real biological system, Fröhlich supposed a biochemical molecule with a linear dimension of 100 Å supporting long wave elastic vibrations leading to frequencies of the order of 10^{11} Hz.[31] Action spectra of microwave-induced biological effects provide support for this estimate.[40]

But Fröhlich has emphasized that both higher and lower frequencies may be involved. "Thus larger units such as DNA-protein complexes might well possess lower frequencies. Higher ones, on the other hand, may be based on a combination of the various rotational and vibrational subgroups of relevant molecules," he comments.[31]

It is interesting to observe that the amide-I vibration in protein that Davydov proposed as the basis for his soliton model[4,7] is precisely a "vibrational subgroup of a relevant molecule." Upon adding dissipation and a source of input metabolic energy to (15), one arrives at a system that is very close to Fröhlich's original concept.[39] From this perspective, the Davydov and Fröhlich theories appear as complementary (rather than competing) explanations for the mystery of energy storage and transport in biological molecules.

ACKNOWLEDGMENTS

It is a pleasure to thank L. MacNeil for calculating Fig. 2, E. Gratton for providing Fig. 3, and H. M. Sobel for Fig. 4.

REFERENCES

1. L. Landau, *Phys. Zeit. Sowjetunion* **3**, 664 (1933).
2. N. J. Zabusky and M. D. Kruskal, *Phys. Rev. Lett.* **15**, 240 (1965).
3. D. Green, *Science* 181, 583 (1973); *Ann. N.Y. Acad. Sci.* **227**, 6 (1974).
4. A. S. Davydov and N. I. Kislukha, *Phys. Stat. Sol.* (b) **59**, 465 (1973); A. S. Davydov, *J. Theor. Biol.* **38**, 559 (1973).
5. V. A. Kuprievich and Z. G. Kudritskaya, preprints #ITP 82-62E, 82-63E, and 82-64E, Institute for Theoretical Physics, Kiev (1982).
6. G. Careri, U. Buontempo, F. Carta, E.Gratton and A. C.Scott, *Phys. Rev. Lett.* **51**, 304 (1983); and G. Careri, U. Buontempo, F. Galluzzi, A. C. Scott, E. Gratton and E. Shyamsunder, *Phys. Rev.* (submitted).
7. A. S. Davydov, *Physica Scripta* **20**, 387 (1979); A. S. Davydov, *Biology and Quantum Mechanics*, Pergamon Press; and *Sov. Phys. Usp.* **25** (12), 899 (1982) and references therein.
8. J. M. Hyman, D. W. McLaughlin and A. C. Scott, *Physica* **3D**, 23 (1981); A. C. Scott, *Phys. Rev. A* 26, 578 (1982); *ibib.* **27**, 2767 (1983); A. C. Scott, *Physica Scripta* **25**, 651 (1982); L. MacNeil and A. C. Scott, *Physica Scripta* (in press).
9. J. A. McCammon, B. R. Gelin and M. Karplus, *Nature* **267**, 585 (1977); M. Levitt, *Nature* **294**, 379 (1981).
10. A. C. Scott, *Phys. Lett.* **94A**, 193 (1983).
11. G. Careri, in *Cooperative Phenomena*, ed. H. Haken and M. Wagner (Berlin: Springer Verlag, 1973), p. 391.
12. Y. N. Chirgadze and N. A. Nevskaya, *Dok. Akad.Nauk SSSR* **208**, 447 (1973).
13. V. P. Gerasimov, *Opt. Spectroscop.* **43**, 417 (1978).
14. J. Frenkel and T. Kontorova, *J. of Phys. (USSR)* **1**, 137 (1939).
15. G. Maret, R. Oldenbourg, G. Winterling, K. Dransfeld and A. Rupprecht, *Colloid and Polym. Sci.* **257**, 1017 (1979).
16. W. N. Mei, M. Kohli, E. W. Prohofsky, and L. L. Van Zandt, *Biopolymers* **20**, 833 (1981); L. L. Van Zandt, K. C. Lu and E. W. Prohofsky, *Biopolymers* **16**, 2481 (1977); and J. M. Eyster and E. W. Prohofsky, *Biopolymers* **13**, 2505 (1974); E. W. Prohofsky, *CMC* **2**, 65 (1983).
17. S. W. Englander, N. R. Kallenbach, A. J. Heeger, J. A. Krumhansl, and S. Litwin, *Proc. Nat. Acad. Sci. (USA)* **77**, 7222 (1980); S. W. Englander, *CMCB* **1**, 15 (1980).
18. S. Yomosa, *Phys. Rev. A*, **27**, 2120 (1983); S. Takeno and S. Homma, *Prog. Theor. Phys.* **70**, 308 (1983).
19. B. F. Putnam, L. L. Van Zandt, E. W. Prohofsky and W. N. Mei, *Biophys. J.* **35**, 271 (1981).
20. S. M. Lindsay and J. Powell, *Biopolymers* (in press).
21. J. A. Krumhansl and J. R. Schrieffer, *Phys. Rev. B* **11**, 3535 (1975).
22. T. R. Koehler, A. R. Bishop, J. A. Krumhansl, and J. R. Schrieffer, *Solid State Comm.* **17**, 1515 (1975).

23. J. A. Krumhansl and D. M. Alexander, in *Structure and Dynamics: Nucleic Acids and Proteins*, eds. E. Clementi and R. H. Sarma (Adenine Press, 1983), p. 61.
24. H. M. Sobell, in *Structure of Biological Macromolecules and Assemblies, Vol. II*, eds. F. Jurnak and A. McPhersonn (New York: Wiley, 1984), (to appear); A. Banerjee and H. M. Sobell, *J. Biomol. Structure and Dynamics 1* (in press).
25. M. Levitt, *Cold Spring Harb. Symp. Quant. Biol.* **47**, 251 (1983).
26. S. Pekar, *Jour. Phys. U.S.S.R.* **10**, 341 (1946); *ibid*, 347.
27. H. Fröhlich, H. Pelzer and S. Zienau, *Phil. Mag.* **41**, 221 (1950); H. Fröhlich, *Adv. in Phys.* **3**, 325 (1954).
28. T. Holstein, *Ann. Phys.* **8**, 325 (1959); *ibid*, 343.
29. S. Yomosa, *J. Phys. Soc. Japan* **52**, 1866 (1983).
30. H. Fröhlich, *Nature* **228**, 1093 (1970); *Jour. Coll. Phen.* **1**, 101 (1973).
31. H. Fröhlich, *Riv. del Nuovo Cim.* **7**, 399 (1977).
32. H. Bilz, H. Büttner and H. Fröhlich, *Z. Naturforsch.* **36b**, 208 (1981).
33. S. Takeno, *Prog. Theor. Phys.* **69**, 1798 (1983); *ibid.* **71**, (1984).
34. M. v. Volkenstein, *J. Theor. Biol.* **34**, 193 (1972).
35. D. E. Green and S. Ji, *Proc. Nat. Acad. Sci. (USA)* **69(3)**, 726 (1972).
36. S. Ji, *Proceedings of the Second International Seminar on the Living State*, ed. R. K. Mishra, held in Bhopal, India (November, 1983) (in press).
37. P. Mitchell, *Eur. J. Biochem.* **95**, 1 (1979).
38. G. Kemeny and I. M. Goklany, *J. Theor. Biol* **40**, 107 (1973); *ibid.* **48**, 23 (1974); G. Kemery, *ibid.* **48**, 231 (1974).
39. H. Fröhlich, *Phys. Lett.* **26A**, 402 (1968); *Int. J. Quant. Chem.* **2**, 641 (1968).
40. S. J. Webb and A. D. Booth, *Nature* **222**, 1199 (1969); Devyatkov et al., *Sov. Phys. Usp.* **16**, 568 (1974); Berteaud et al., *C. R. Acad. Sci. Paris* **281D**, 843 (1975); W. Grundler, F. Keilmann and H. Fröhlich, *Phys. Lett.* **62A**, 463 (1977); W. Grundler and F. Keilmann, *Phys. Rev. Lett* **51**, 1214 (1983).

BOOK LIST

K. Lonngren and A. C. Scott, eds. *Solitons in Action*. New York: Academic Press, 1978.

A. R. Bishop and T. Schneider, eds. *Solitons and Condensed Matter Physics*. Berlin: Springer-Verlag, 1978.

F. Calogero. *Nonlinear Evolution Equations Solvable by the Spectral Transform*. London: Pitman, 1978.

V. E. Zakharov, S. V. Manakov, S. P. Novikov and L. P. Pitayevsky. *Theory of Solitons: The Method of the Inverse Scattering Problem*. Moscow: Nauka, 1980 (in Russian).

G. L. Lamb. *Elements of Soliton Theory*. New York: John Wiley, 1980.

R. K. Bullough and P. J. Caudrey, eds. *Solitons*. Berlin: Springer-Verlag, 1980.

G. Eilenberger. *Solitons*. Berlin: Springer-Verlag, 1981.

W. Eckhaus and A. V. Harten. *The Inverse Scattering Transformation and the Theory of Solitons*. Amsterdam: North-Holland, 1981.

M. J. Ablowitz and H. Segur. *Solitons and the Inverse Scattering Transform*. Philadelphia: SIAM, 1981.

F. Calogero and A. Degasperis. *Spectral Transform and Solitons*. Amsterdam: North-Holland, 1982.

R. K. Dodd, J. C. Eilbeck, J. D. Gibbon and H. C. Morris. *Solitons and Nonlinear Wave Equations*. London: Academic Press, 1982.

P. G. Drazin. *Solitons*. New York: Cambridge University Press, 1983.

THEODORE T. PUCK
Director, Eleanor Roosevelt Institute for Cancer Research, and
Professor of Biochemistry, Biophysics and Genetics, and Professor of Medicine,
University of Colorado Health Sciences Center

The New Biology and Its Human Implications

SUMMARY OF REMARKS

The original DNA revolution indicated the nature of molecular storage of information in the cell and how this information is to be incorporated into specific protein molecules so that understanding was achieved of how the structural elements and machinery of the simplest living cell could be constructed.

The next step, which is currently taking place, involves extending this understanding of enormously more complex organisms, which are characterized by possession of the phenomenon of differentiation.

Somatic cell genetics was invented to acquire the necessary information about the genetic structure and function of the complex organisms like the mammals. Together with the new approaches of recombinant DNA technology and new methods for analysis of the patterns of protein biosynthesis in the complex mammalian cells, many new levels of understanding are being achieved. At present, less than 1/10 of 1% of the individual biochemical pathways which make up the body's metabolic chains have been identified. New general methods are now available which promise to unlock many or most of these critical pathways.

For the first time in history, a framework of conceptual understanding is being built to support and greatly extend the still largely empirical discipline of medicine. New promise for revolutionizing food supplies of the world appear at hand. Finally, application of cellular genetics and molecular biology to the nervous system is opening up new approaches to understanding of the mind.

These new powers could, if appropriately used, bring a new era of health and fulfillment to mankind. Scientists must see that this message reaches the peoples and governments of the planet.

ACKNOWLEDGMENT

This work acknowledges grants from Lucille P. Markey Charitable Trust, the R. J. Reynolds Industries and National Institutes of Health H.D. 02080.

HANS FRAUENFELDER
Department of Physics, University of Illinois at Urbana-Champaign

Biomolecules

The study of the structure and function of biomolecules connects biology, biochemistry, chemistry, and physics. I believe that major progress and a deep understanding of these complex systems will only be possible in a truly interdisciplinary collaboration. Of course, a chemist will look at proteins differently from a biologist, and a physicist's interest may again be different. My own approach is best characterized by a dinner conversation. Some years ago I had the good fortune of joining Stan Ulam for dinner at that famous culinary pinnacle, the Los Alamos Inn. I described my work to Stan who then remarked: "Aha, ask not what physics can do for biology, ask what biology can do for physics." It may indeed be possible that biomolecules will yield results of interest to physics. On the one hand, biomolecules are truly complex. As I will describe later, proteins have highly degenerate ground states. If we define complexity, κ, as the logarithm of the number of "components,"[1] proteins have $\kappa \gg 1$. Thus, while the complexity is far smaller than that of the brain or of a sociological system, it is large enough to lead to nontrivial results. Moreover, in contrast to glasses and spin glasses (at least the experimental ones), proteins are tailor-made, all proteins of a given type from a given system have exactly the same number of constituents, and experimental work can be performed on very well-defined entities.

1. PROTEINS

For the following discussions, a few general facts about proteins are necessary. I will only describe rudimentary aspects[2]; details can be found in a number of texts that can even be read by physicists.[3,4]

Proteins are the machines of life. They are constructed from twenty different building blocks, the amino acids. As indicated in Fig. 1, of the order of 100 amino acids are covalently linked to form a long linear chain. The arrangement of the amino acids in this chain, the *primary structure*, determines the final *tertiary structure* and the function of a particular protein. In a proper solvent, the linear chain will fold into the space-filling tertiary structure, the working protein. The final protein looks like a miniature crystal, consisting of the order of 1000 atoms and with linear dimensions of the order of a few nanometers.

The illustrations in textbooks make proteins appear as rigid structures. A closer look at the structure and function of a very important protein, hemoglobin, makes it clear, however, that motion is important. Hemoglobin transports oxygen and the dioxygen molecule is stored inside the hemoglobin during the ride from the lung to wherever O_2 is used. X-ray diffraction data indicate that there is no open path from the outside to the storage site. If hemoglobin were rigid, it could not fulfill its function.

A second look at Fig. 1 shows why proteins are flexible and can behave like machines rather than like pieces of rock. The forces along the protein backbone, the polypeptide chain, are "strong" (covalent) and cannot be broken by thermal fluctuations. The forces that hold the tertiary structure together are "weak," mainly hydrogen bonds and van der Waals forces. These weak bonds are continuously broken and reformed: the protein breathes. This breathing motion is essential for many functions.

2. A PROTEIN IN ACTION

In order to study the states and motions of a protein, we must look at a protein in action. For the past decade, we have been investigating a very simple process, the binding of a small molecule (ligand) such as dioxygen (O_2) or carbon monoxide (CO) to myoglobin.[5-7] Myoglobin is a protein of molecular weight 17.9 kD, with dimensions $2.5 \times 4.4 \times 4.4$ nm^3, that reversibly stores dioxygen in muscles.[2,3] An approximate cross-section is shown in Fig. 2a. Embedded in the protein matrix is a planar organic molecule, heme, which contains an iron atom at its center. Storage of O_2 or CO occurs through covalent binding of the small molecule at the iron atom. We can look at the association and dissociation of the ligand (say, CO) in a number of ways.

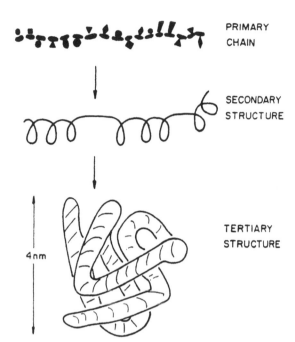

PRIMARY
CHAIN

SECONDARY
STRUCTURE

TERTIARY
STRUCTURE

4 nm

FIGURE 1 The linear polypeptide chain (primary sequence) folds into the final
tertiary structure.

(i) THE BINDING IN REALITY. A CO molecule in the solvent around Mb
executes a Brownian motion in the solvent, moves into the protein matrix, migrates
through the matrix to the heme pocket B, and finally binds covalently to the heme
iron (Fig. 2a). The binding process can be studied with many different tools.[8] We use
flash photolysis: a sample of liganded proteins (MbCO) is placed into a cryostat. At
the proper temperature, the sample is hit with a laser pulse which breaks the Fe-CO
bond: $MbCO + h\omega \rightarrow Mb + CO$. The rebinding, $Mb + CO \rightarrow MbCO$, is followed
optically. Studies of rebinding over a wide range of time and temperature suggest
that binding follows the pathway indicated in Fig. 2a. The theoretical treatment of
the binding process is difficult. In physics, progress in describing phenomena often
starts with models that describe some aspects well, but totally miss others (the
single-particle shell model and the collective model in nuclear physics). Ultimately,
a unified model incorporates the essential aspects of the early attempts. Progress
in models for protein dynamics may follow a similar path.

(ii) THE SINGLE-PARTICLE (STATIC) MODEL. In the simplest model, we
assume that the protein forms an effective static potential in which the CO molecule
moves. The experiments imply that the potential is as sketched in Fig. 2b, where
S represents the solvent, M the protein matrix, B the heme pocket, and A the
covalent binding site at the heme iron. The general behavior in such a potential
is easy to describe, but a full quantitative treatment is difficult and has not yet
been achieved. The ligand will perform a complicated random walk in the potential

and entropy (the number of states as function of the reaction coordinate) plays an important role.

(iii) PROTEIN MOTIONS. The single-particle model is static and does not explicitly consider the motions of the protein. We know, however, that the protein motion is important: The total "binding energy" of a protein, defined as the difference in Gibbs energy between the folded and unfolded state (Fig. 1) is of the order of 1 eV and hence very small. A ligand moving through the protein matrix will affect the protein strongly and the covalent binding, which also involves an energy of the order

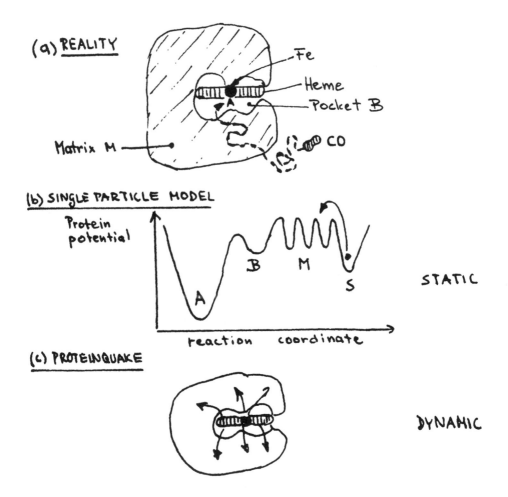

FIGURE 2   A protein process, ligand binding, and two extreme models.

of 1 eV, causes a major perturbation of the protein structure ("proteinquake"). As second model (Fig. 2c), we therefore neglect the motion of the CO and only consider the protein motions that are caused by the formation or breaking of the covalent bond at the iron. We will return to this problem.

(iv) UNIFIED MODEL. In a complete model, the motions of the protein and the ligand would be considered together. At present, not enough is known about the dynamics of the protein and the interaction of the ligand with the protein to formulate such a model.

3. ENDLESS PROCESSES

The observation of the binding of CO to Mb at temperatures below about 200 K yielded a result that was at first very surprising.[5,9] Below 200 K, the CO molecule remains in the heme pocket after photodissociation and rebinds from there. The rebinding process, denoted by I, is then "geminate" or intramolecular. This feature permits a detailed study of the mechanism of the formation of the Fe-CO bond.[10] The signal property of process I is its time dependence, shown in Fig. 3. Process I is not exponential in time, but can be approximated by a power law,

$$N(t) = (1 + t/t_o)^{-n}. \tag{1}$$

Here, t_o and n are temperature-dependent parameters. For the binding of CO to Mb between 60 and 160 K, n is approximately given by

$$n = 2.8 \times 10^{-3} \, T/K. \tag{2}$$

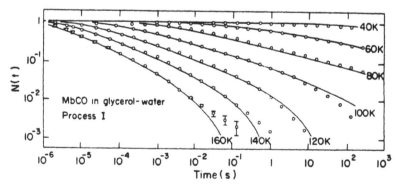

FIGURE 3   Time dependence of the binding of CO to Mb between 40 and 160 K. $N(t)$ is a fraction of Mb molecules that have not rebonded a CO molecule at the time t after photodissociation. (After reference 5.)

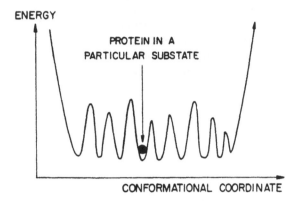

FIGURE 4   Protein energy versus configurational co-ordinate. A large number of conformation substates (potential minima) have essentially the same energy. The dot represents a protein in a particular substate.

A "fractal" time dependence was first observed by Weber in 1835[11] and the fascinating history is sketched by Bendler.[12] Processes as described by Eqs. (1) and (2) have been called "endless" and the reason is clear. At 60 K, $n = 0.17$ and $t_o = 0.16\ s$ for MbCO. Rebinding becomes observable at about 10^{-3} s, but the time required for the rebinding of 99.9% of the Mb molecules is 2×10^9 years. Nonexponential behavior is observed in many different systems, but why do proteins exhibit it?

The simplest explanation that we have been able to find is as follows. The covalent binding $B \rightarrow A$ at the heme iron involves a potential barrier that must be overcome. If the barrier is the same in all Mb molecules, binding will be exponential in time. If, however, different Mb molecules have different barriers, the nonexponential time dependence can be described easily.[9,5,2] But why should different Mb molecules have different activation barriers?

The crucial point is that a protein cannot be in a unique state of lowest energy, its ground state is highly degenerate. In a given state (say, MbCO) the protein can assume a large number of conformational substates.[13] All substates perform the same function, but differ slightly in the geometrical arrangement of the atoms. In a one-dimensional abstraction, we can represent the energy of a protein as shown in Fig. 4 as a function of a conformational coordinate. At low temperatures, each protein will have a distribution of barrier heights and consequently show a nonexponential binding as in Fig. 3. At high temperatures (300 K), transitions among the substates occur and a given protein will move from substate to substate. If the ensemble is studied with a technique characterized by a time much longer than required for transitions among substates, the protein ensemble appears homogeneous.

4. PROTEIN MOTIONS

The existence of substates leads to a separation of protein motions into two classes, equilibrium fluctuations (EF) and functionally important motions (fims). Fig. 5

gives the basic ideas. A given protein can exist in a number of states, for instance, MbCO and deoxyMb. Each of these states contains a large number of conformational substates, denoted by CS. Transitions among the substates are equilibrium fluctuations. The protein action, the transition from MbCO to deoxyMb or from deoxyMb to MbCO, is performed through fims. In order to fully understand the dynamic connection between protein structure and function, both are connected through fluctuation-dissipation theorems.[14-18]

While equilibrium fluctuations can be studied on resting proteins, the exploration of fims must involve proteins in action. When asked for a title for a talk at a birthday symposium for David Pines, I intended to make a joke and suggested "Do Proteins Quake?". During the preparation of the talk, I realized that the joke was on me—proteins indeed do quake, and the investigation of proteinquakes can yield considerable insight into protein dynamics.[19,20] Consider first an earthquake, as shown in Fig. 6a. In some regions of the world, for instance near SLAC, stress builds up. When the stress exceeds a critical value, it is relieved through an earthquake. The quake results in the propagation of waves and of a deformation. In a protein, events are similar: a stress is created at the site of a reaction. Consider for instance the photodissociation of MbCO as in Fig. 6b. Before the laser flash, the entire protein is in the liganded conformation. Immediately after photodissociation, the heme and the protein are still in the liganded structure, but now far off equilibrium. Return to equilibrium occurs through a proteinquake: the released strain energy is dissipated through waves and through the propagation of a deformation. The proteinquake can be followed through observation of suitable markers in visible, near-infrared, and resonance Raman spectra.

The main results of various experiments taken together[20] indicate that the proteinquake occurs in a series of steps. So far, four distinct fims have been recognized

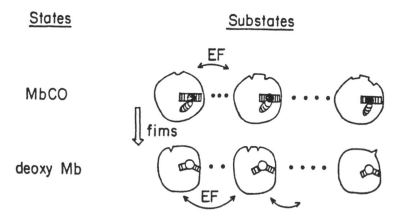

FIGURE 5   States, substates, equilibrium fluctuations, and fims.

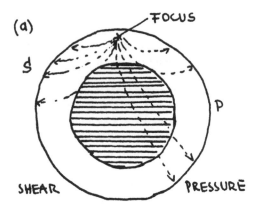

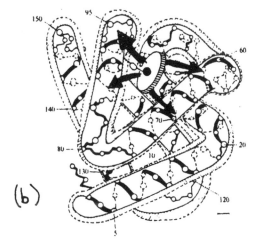

FIGURE 6 (a) Earthquake; (b) Proteinquake. The focuses of the proteinquake are at the heme iron. Also shown in the figure is the breathing motion of the protein: the shaded area gives the region reached by the backbone because of fluctuations (after ref. 13).

and some of their properties established. The relation between dissipation (fims) and fluctuations (EF) suggests that the four fims should be complemented by four types of equilibrium fluctuations. These, in turn, imply four tiers of substates. We consequently arrive at a model for the structure of myoglobin as indicated in Fig. 7.

Figure 7 shows that proteins have a hierarchical structure, and thus suggests a close similarity between proteins and glasses.[21-25]

These results are clearly tentative and all aspects remain to be explored in much more detail. They indicate, however, that a close interaction among biologists, biochemists, and theoretical and experimental physicists is necessary for progress.

In a minor way, the work that I have described already involves an I^2N, an "Interdisciplinary International Network," as indicated in Fig. 8. The collaboration is necessary because no single individual knows all aspects, no single group possesses all the required tools and techniques, and no single protein can yield all the information needed for a deep understanding.

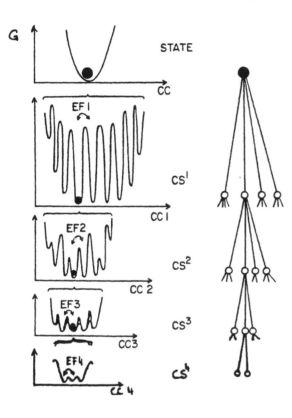

FIGURE 7 The hierarchical arrangements of substates in myoglobin. Four tiers of substates are believed to exist.

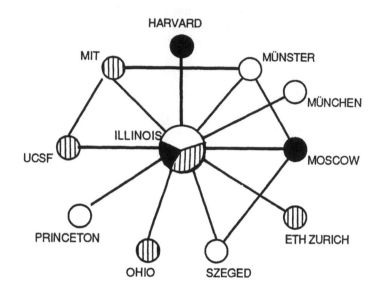

FIGURE 8 I^2N active in the exploration of protein dynamics.

REFERENCES

1. R. G. Palmer, *Adv. Physics* **31**, 669 (1982).
2. H. Frauenfelder, *Helv. Phys. Acta* **57**, 165 (1984).
3. R. E. Dickerson and I. Geis, *Hemoglobin* (Menlo Park, CA: Benjamin/Cummings, 1983).
4. L. Stryer, *Biochemistry* (San Francisco, CA: W. H. Freeman, 1981).
5. R. H. Austin, K. W. Beeson, L. Eisenstein, H. Frauenfelder, and I. C. Gunsalus, *Biochemistry* **14**, 5355 (1975).
6. D. Dlott, H. Frauenfelder, P. Langer, H. Roder, and E. E. DiIorio, *Proc. Natl. Acad. Sci. USA* **80**, 6239 (1983).
7. H. Frauenfelder and R. D. Young, *Comments Mol. Cell. Biophysics* **3**, 347 (1986).
8. E. Antonini and M. Brunori, *Hemoglobin and Myoglobin in Their Reactions with Ligands* (Amsterdam: North-Holland, 1971).
9. R. H. Austin, K. Beeson, L. Eisenstein, H. Frauenfelder, I. C. Gunsalus, and V. P. Marshall, *Phys. Rev. Letters* **32**, 403 (1974).
10. H. Frauenfelder and P. G. Wolynes, *Science* **229**, 337 (1985).
11. W. Weber, *Götting. Gel. Anz.* (1835), p. 8; *Annalen der Physik und Chemie* (Poggendorf) **34**, 247 (1835).
12. J. T. Bendler, *J. Stat. Phys.* **36**, 625 (1984).
13. H. Frauenfelder, G. A. Petsko, and D. Tsernoglou, *Nature* **280**, 558 (1979).
14. H. B. Callen and T. A. Welton, *Phys. Rev.* **83**, (1951).
15. R. Kubo, *Rep. Progr. Phys.* **29**, 255 (1966).
16. M. Suzuki, *Progr. Theor. Phys.* **56**, 77 (1976).
17. M. Lax, *Rev. Mod. Phys.* **32**, 5 (1960).
18. F. Schlögl, *Z. Physik* **B33**, 199 (1979).
19. H. Frauenfelder, "Ligand Binding and Protein Dynamics," *Structure and Motion: Membranes, Nucleic Acids, and Proteins*, Eds. E. Clementi, G. Corongiu, M. H. Sarma, and R. H. Sarma (Guilderland, NY: Adenine Press, 1985).
20. A. Ansari, J. Berendzen, S. F. Bowne, H. Frauenfelder, I. E. T. Iben, T. E. Sauke, E. Shyamsunder, and R. D. Young, *Proc. Natl. Acad. Sci. USA* **82**, 5000 (1985).
21. D. Stein, *Proc. Natl. Acad. Sci. USA* **82**, 3670 (1985).
22. G. Toulous, *Helv. Phys. Acta* **57**, 459 (1984).
23. M. Mézard, G. Parisi, N. Sourlas, G. Toulouse, and V. Virasoro, *Phys. Rev. Lett.* **52**, 1156 (1984).
24. R. G. Palmer, D. L. Stein, E. Abrahams, and P. W. Anderson, *Phys. Rev. Lett.* **53**, 958 (1984).
25. B. A. Huberman and M. Kerzberg, *J. Physics A* **A18**, L331 (1985).

B. A. HUBERMAN
Xerox Palo Alto Research Center, Palo Alto, CA 94304

Computing With Attractors:
From Self-repairing Computers, to
Ultradiffusion, and the Application of
Dynamical Systems to Human Behavior

It is seldom that one has the opportunity of spending such a pleasant weekend in the company of scholars from so many fields. Even more remarkable is the fact that, although different in outlook and methodologies, the presentations at this workshop display a serious attempt at bridging the gap that separate our disciplines. From reports of studies of the origin of life, to attempts at understanding the nature of daydreaming, one perceives the great potential that interdisciplinary approaches might have for the solution of these problems.

My talk will mostly deal with the convergence of two apparently dissimilar disciplines, dynamical systems and computers, and its implications for the understanding of both complexity and biological computation. Before closing, I will also mention some speculations about the application of dynamical systems to human behavior. Since it is superfluous to remind you of the phenomenal progress that we are witnessing in computer technology, I will start by giving a very short status report on our current understanding of the dynamics of nonlinear systems, and of the problems which I perceive lie ahead.

A. COMPLEX SYSTEMS

The past few years have witnessed an explosive growth in the application of nonlinear dynamics to physical and chemical systems.[1] In particular, the recognition that very simple dissipative, deterministic classical systems can display chaos, has led to a new approach to problems where erratic, noisy behavior seems to be prevalent. With this new paradigm in our midst, we now feel confident that we have the needed tools for analyzing the dynamics of nonlinear systems, *provided* they appear to an external probe as low dimensional in their phase spaces.

Beyond this fairly placid scenario lies the terra incognita of more complex systems and their associated dynamics. Here, one is dealing with dimensionalities such that neither simple geometry nor statistical mechanics can be put to good use. And yet, complexity is pervasive and full of interesting properties. Structures such as living organisms and computers are examples of systems displaying self-organizing properties and non-trivial dynamics which at present defy analytical understanding. If a coherent picture of their behavior is to emerge, it will have to be based on both new theories and crisp data produced by controlled experiments on systems which encapsulate the essence of complexity. Moreover, these studies will hopefully lead to a sharpening of the concept of complex system. Presently, the word complexity itself seems to mean different things to different people and, in spite of the existence of mathematical tools such as algorithmic complexity and entropy, we still lack a precise definition of such an important notion. For the time being, I will use complexity in its simplest form, i.e., as conveying the idea that a system as a whole is more than the sum of its parts, and that its behavior is non-trivial to describe. I will also show below how the appearance of some hierarchical structure in such systems leads to interesting universal dynamics.

B. EMERGENCE OF COMPUTATIONAL BEHAVIOR

Parallel computing structures, which are common in nature, provide an ideal experimental tool when implemented in actual machines. This allows for a detailed analysis of the dynamics of highly concurrent processes which are not often experimentally accessible in the real world. By performing quantitative experiments on them, one hopes to both uncover new phenomena and to abstract general laws governing their behavior. Typical questions that can be asked are about self-organization and its dynamics, adaptation, and the range of behavioral functions of the brain that can be reproduced by the collective behavior arrays of simple, locally connected, computing elements. Answers to these questions[2,3] are important in understanding the emergence of complex behavior out of a collection of simple units, in determining to what extent VLSI structures can be made to behave in adaptive fashion and, more generally, in elucidating the global behavior of systems made up of elementary computational cells.

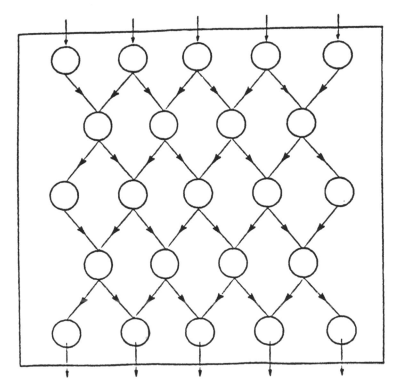

FIGURE 1   Diagram of a typical array.

Our experimental approach to these issues considers arrays of simple local units that exhibit some interesting property. A typical architecture is schematically shown in Figure 1. Each processor operates on integer data received locally from its neighbors. Overall input and output to the machine takes place only along the edges and the computation is systolically advanced from row to row in step with a global clock. Each processor has an internal state, represented by an integer, which can only take on a small set of values depending on a given adaptive rule. The unit determines its local input based on its inputs and its internal state. At each time step, every element receives data values from the units to its upper left and right and computes its output, which is then sent to its neighbors.

For various values of the array parameters, we then quantitatively examine the emergence of global computational behavior as a function of time. Within this context, we have recently shown that there is a class of architectures that can be made to compute in a distributed, deterministic, self-repairing fashion, by exploiting the existence of attractors in their phase spaces.[4] Such a mechanism leads to computing structures which are able to reliably learn several inputs and to recognize them even when slightly distorted. In the language of dynamical systems, this corresponds to the appearance of fixed points in the phase space of the system. Furthermore, the contraction of volumes in phase space makes these fixed points attractive in the sense that perturbations in either data or the state of the array quickly relax back to the original values. The set of inputs which map into a given output defines the basin of attraction for that output, as illustrated in Figure 2a.

Since there are many such basins of attraction, a natural question concerns the possibility of changing them at will with local rules. In other words, one is interested in dynamically modifying the basins of attraction in order to include or exclude a particular set of inputs. Figures 2a and 2b show schematically how this adaptive mechanism works. These new processes of coalescence and dissociation of attractors lead to results analogous to Pavlovian conditioned reflexes. Furthermore, through quantitative measurements of the sizes of the basins of attraction before and after associations, we were able to determine that such

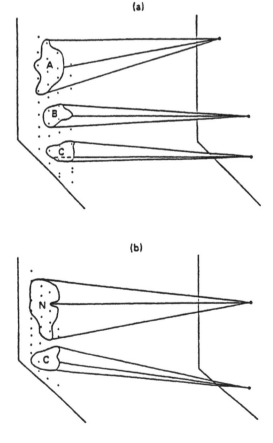

FIGURE 2 (a) Basins of attraction for 3 inputs, [A], [B], and [C] mapping into 3 different outputs. (b) The attractors after coalescence of sets [A], and [B] into a new basin of attraction.

dissociation of attractors lead to results analogous to Pavlovian conditioned reflexes. Furthermore, through quantitative measurements of the sizes of the basins of attraction before and after associations, we were able to determine that such arrays exhibit generalization and differentiation of inputs in their behavior. Besides their intrinsic value, these results open the door to exciting applications of this computing scheme to both pattern and speech recognition.

C. BIOLOGICAL COMPUTATION

I would now like to say a few words about the possible relationship between this new paradigm and biological computations.[5] In doing so, I am aware of the case with which one should extrapolate computational results into neurobiology. Just as learning about birds' flight cannot be totally accomplished by designing airplanes, to have computing structures which mimic some aspects of brain behavior does not imply we know how brains work. This, in turn, raises the still unclear issue of the role of simulations in describing reality, a problem which is bound to grow as computers become an increasing source of experimental data in the sciences.

Information processing in biological systems possesses a number of intriguing fundamental characteristics that are difficult to understand in terms of sequential computing mechanisms. One of these is fault tolerant behavior, both with respect to internal failures and to input data errors, allowing the system to operate with distorted, or fuzzy, inputs. Further examples are associative memory and conditional learning, where the ability to learn something is facilitated by previous knowledge of something similar to it. Finally, there is the ability to adapt to changes in the environment, or plasticity, with the associated mechanism of selection out of degenerate initial configurations.

Our results suggest the hypothesis that the brain might operate reliably, even though individual components may intermittently fail, by computing with dynamic attractors. Specifically, such a mechanism exploits collective behavior of a dynamical system with attractive fixed points in its phase space.

Although its applicability to biological systems cannot be proven, the study of such computing structures suggests that our hypothesis may, indeed, be relevant to biological systems regardless of the detailed operation of individual neurons. It also produces a deterministic alternative to models of the brain based on probabilistic neural nets.[6]

D. ATTRACTORS ON DISCRETE STRUCTURES

The phenomenon of deterministic computation with attractive fixed points leads naturally to a mathematical description of such processes using attractors on discrete structures. Since the dynamics of a quantity that takes on a finite set of values and changes only at discrete instants of time is governed by discrete maps, the time evolution of such computing arrays will generate contractive mappings of a finite set into itself. This new description, which is now being developed, contains an interesting mix of combinatorial analysis and dynamics, and is bound to give us new vistas on the subject of dynamical systems. Moreover, such a theory should contain enough predictive power so as to tell us which general computational rules and architectures will perform given functions.

E. ULTRADIFFUSION

Complexity in natural and artificial systems often manifests itself in hierarchical fashion: at any given level of the system, the effect of the lower echelons can, for all practical purposes, be integrated over while the larger scale structures are essentially frozen and act as static constraints. This architecture of organization appears in economic systems, formal research organizations and computing structures, and produces nonergodic behavior in many problems with a hierarchy of energy barriers. Molecular diffusion in complex macromolecules, and spin glasses, provide examples where this behavior is found.

A common feature of hierarchical systems is that they can be characterized by an ultrametric topology, i.e., a distance can be defined so that any triplet of points can be labelled in such a way that their respective distances form an isosceles triangle. For such topologies, to determine the time evolution of stochastic processes entails solving the dynamics of Markovian matrices which are near decomposable, a problem which was posed many years ago by Simon and Ando in their study of the aggregation of economic variables.[7]

In order to solve for the dynamics of hierarchical systems, we have recently constructed a simple one-dimensional model which possesses such a topology and studied its dynamical properties by explicit renormalization.[8] We showed that the relaxation of the autocorrelation function obeys a universal algebraic law which we termed ultradiffusion. In particular, for thermally activated processes, its long-time behavior is characterized by an effective dimensionality which is temperature dependent, leading to an anomalous low-frequency spectrum reminiscent of the $1/f$ noise type of phenomena observed in a variety of systems.

One interesting aspect of these results is given by the fact that such ultrametric topology also appears in probabilistic computing schemes which have been advanced over the years to model neural nets.[9,10] These models, which appeal to the analyogy between Hebbian synapses and Ising spins, contain ingredients similar to the spin glass problem and should, therefore, exhibit slow decays of their autocorrelation functions. I should mention that because of their stochastic nature, such lack of convergence poses serious problems when trying to use them as computing structures, since the existence of a hierarchy of energy barriers implies that it is not clear when to halt a given computation.

F. DYNAMICAL SYSTEMS AND HUMAN BEHAVIOR

I would like to finish this presentation in a more speculative note and report some studies that we have undertaken to characterize the dynamics of behavior. Although unrelated to computation, these issues will illustrate how a very complex system such as the brain can sometimes be phenomenologically described using techniques which invoke only a few variables. Moreover, since part of the problem in understanding human behavior is a development of a language in which to codify it, we have tried to show how the tools of dynamical systems theory can be used to study its unfolding.[11]

Experimental work in neurobiology demonstrates a boggling array of complex neurophysiological and neurochemical interactions. The existence of many neuro-transmitters and the complications of synaptic function create intricate paths for neuronal activity. Nevertheless, the dynamical modelling of such systems is aided by three important points:

1. The imprecision in the observation of human psychopathology. As a consequence, one is likely to see on any time scale only the broad distinctions between fixed points, limit cycles and chaos.
2. The substantial delays between the reception of a neurochemical process and its ultimate physiological effect.
3. The wide temporal separation between many neurophysiological processes in a given system. These range from fractions of a second for certain GABA receptors, to intervals of minutes for short-term desensitization of adrenergic activity.

The first two properties justify the use of differential delay equations in simulating those systems, while the last one offers a methodology for constructing dynamics over many time scales. Within this context, we have recently studied the dynamics of a model of the central dopaminergic neuronal system.[12] In particular, we showed that for certain values of a parameter which conrols the efficacy of dopamine at the postsynaptic receptor, chaotic solutions of the equations appear. This prediction correlates with the observed increased variability in behavior among schizophrenics and the rapid fluctuations in motor activity among Parkinsonian patients chronically treated with L-dopa.

These results, which still have to survive the scrutiny of controlled experiments, suggest that the impact of nonlinear dynamics will be felt on fields far removed from its original concerns, and in so doing, it will itself undergo unforeseen changes. Thank you.

ACKNOWLEDGEMENTS

It is a pleasure to thank Tad Hogg and Michel Kerszberg for their collaboration and our endless conversations. Thanks to their patience and enthusiasm, our shared vision of an emerging field became concrete enough so as to produce new results. I also thank Roy King for the many discussions that led to our work on dynamics of human behavior. Part of this research was supported by ONR contract N00014-82-0699.

REFERENCES

1. See, for instance, *Dynamical Systems and Chaos*, Proceeding of the Sitges Conference, Springer Lecture Notes in Physics (1982).
2. G. M. Edelman and V. B. Mountcastle, *The Mindful Brain* (Cambridge, MA: MIT Press, 1978).
3. J. von Neumann, in *Theory of Self-Reproducing Automata*, ed. A. W. Burks (Urbana, IL: University of Illinois Press, 1966).
4. B. A. Huberman and T. Hogg, *Phys. Rev. Lett.* **52**, 1048 (1984).
5. T. Hogg and B. A. Huberman, *Proc. Natl. Acad. Sci. (USA)* **81**, 6871 (1984).
6. See, for example, M. Rochester, J. H. Holland, L. H. Haibt, and W. L. Duda, *IRE Trans. Inf. Theory* **IT-2**, 80 (1956); and G. L. Shaw and K. J. Roney, *Phys. Lett* **74A**, 146 (1979).
7. H. A. Simon and A. Ando, *Econometrica*, **29**, 111 (1961).
8. B. A. Huberman and M. Kerszberg, *J. Phys. A.* **18**, L331 (1985).
9. W. A. Little, *Math. Bio* **19**, 101 (1974); and J. Hopfield, *Proc. Natl. Acad. Sci. USA*, **79**, 2554 (1982).
10. M. Y. Choi and B. A. Huberman, *Phys. Rev.* **A28**, 1204 (1983).
11. R. King, J. D. Barchas, and B. A. Huberman, in *Synergetics of the Brain*, eds. E. Basar, H. Flohr, H. Haken and A. J. Mandell (Springer, 1983), pp. 352–364.
12. R. King, J. D. Barchas, and B. A. Huberman, *Proc. Natl. Acad. Sci. USA*, **81**, 1244 (1984).

FRANK WILCZEK
Institute for Theoretical Physics, University of California, Santa Barbara, CA 93106

Fundamental Physics, Mathematics and Astronomy

I had to make some quite arbitrary decisions as to what I could include and not include, and I could easily imagine someone else very clever or myself in a different mood discussing a rigorously disjoint set of topics. What I chose to do is to talk about three relatively specific, very important problems in physics, astronomy, and to a lesser extent, mathematics and try to generalize from these problems. I chose problems which do have a true interdisciplinary component and, at the same time, are among the most important problems we are currently up against. In each of these, I shall go from the specific to the general, so do not be misled by the headings. They refer to the vague generalities that come at the end.

I. NEW SENSORY SYSTEMS

We now have from microphysical considerations what I think is a very good candidate for a complete model of formation of structure in the universe. More and more definite evidence over the last 20 years has been accumulating for the Bing Bang cosmology, and it is now quite generally established. The new development over the last ten years or so is that we have obtained a much better idea of some unknowns, some of the parameters in the Big Bang cosmology which previously had to be put in as initial conditions with no understanding. We now have real physical insights about what they should be. In particular, we have a strong theoretical prejudice based on reasons which I can not go into in great detail here: that the density of the universe, on the average, should be equal to the critical density. If the universe were slightly more dense than it is, it would eventually collapse. It is just poised on the verge of collapsing. That means the density in terms of Newton's constant G_N and Hubble's parameter H is $3H^2/8\pi G_N$ or numerically 10^{-29} grams per cubic centimeter. Ordinary matter consists of about 10^{24} atoms per cubic centimeter. The density of the universe as a whole is about one atom per cubic meter, a number I find astonishing. Nevertheless, that is approximately the density of the whole, and we want to know whether the universe has that density or slightly less. The theoretical prejudice, which is not contradicted yet by experiments, is that it should be almost precisely equal.

Another key idea concerns the nature of the initial fluctuations. These fluctuations are supposed to be quite small at first, growing by gravitational contraction. There is fairly direct evidence, namely the isotropy to $\gtrsim 10^{-4}$ of the microwave background radiation, that the seed fluctuations were quite small. All the later structure evolves from very small wrinkles early on. These fluctuations are supposed to be fluctuations of mass but not in chemical content. That is called adiabatic fluctuations.

Finally, the only other initial condition that needs to be fixed to make the whole think go as a well-determined problem is the scale and spatial spectrum of fluctuations. What do the fluctuations on different size scales look like? There is a very elegant candidate for the spectrum which emerges from microphysical considerations. As the universe gets older in the Big Bang picture, we get to see more and more of it. There are parts of the universe which are just now becoming visible to us. In other words, the speed of light has just been caught up with the distance from the newly exposed regions in ten billion years; they are ten billion light years away. At earlier times, when the universe was nine billion years old, you could see less. The hypothesis of scale-invariant fluctuations is that the universe fluctuates by very small amounts from critical density at all times and that the amplitude of the fluctuations—the size of the fluctuations—is independent of when you look. If you weight the universe at any time, it should have tiny deviations, $\sim 10^{-4}$, from critical density, and the nature of the deviations should not depend on time.

The immediate consequence of the first of our three principles is that 90% of the mass of the universe has to be in some nonluminous form. To check how much mass there is in the universe, the most naive procedure, of course, is to count up everything you see and estimate how much it weighs. Add all the visible matter up that way—all the things that are in stars, gas clouds and so forth—and you do not get this critical density, but roughly 1/30th of it. Not enough. On the other hand, the direct gravitation experiments—basically trying to verify Kepler's law on a galactic scale—indicate that there are large amounts of nonluminous mass not in the ordinary forms of protons, neutrons and electrons, but in some other form which is nonluminous. That is, you look at objects—gas clouds or individual stars—rotating far from the luminous regions of galaxies where the light of the galaxy has fallen off exponentially, where it is practically zero. If all the mass were also concentrated in this small region similar to the light, then the velocity of these objects should fall like $1/\sqrt{r}$. This is Kepler's law, the same as for planets orbiting the sun. It is found instead that the velocity in many, many galaxies (hundreds have been measured) is constant—does not fall—indicating that the galactic mass is not concentrated where the light is. It is spread out over larger regions, so as you go farther and farther out, you have more and more stuff pulling on you, and you have to move fast to keep from falling in. In this way we know that there is some nonstandard, nonluminous form of matter, which is at least as much matter as the more ordinary visible forms. We need about ten times as much or thirty times as much to satisfy our theoretical expectations. So we have a great scandal, that we do not know what most of the universe is made out of. There are various ideas of particle physics about what it can be. These probably will not be more than names to most of you: axions are a strange sort of particle that can mediate long-range forces and that are being looked for experimentally; photinos are another strange kind of particle associated with low-energy supersymmetry, which are being looked for at accelerators. For both of these you can compute how much there would be of them if they existed. And if you trace out the history of the Bing Bang, you find they could be produced in roughly enough quantity to provide the missing mass. So there is a great discovery to be made here, but we do not know yet what it is.

As I said, those principles I told you, these refinements of the Big Bang, allow us to build up quite a detailed picture of galaxy formation, of the large-scale structure of the universe. However, the predictions we can made are very difficult to test, because after all, when we look at galaxies what we see is debris of an explosion that occurred ten billion years ago. All the stuff that we actually see has been chemically processed and has gone through convective motions, and it is very difficult to reconstruct what exploded from the cinders of the explosion. However, there are some striking qualitative predictions and even semi-quantitative ones which will be tested in the near future by very interesting means, I think. We have to account, first of all, for the main mystery: why the dark matter is not as clustered as the ordinary luminous matter. The most plausible idea about that astrophysically is that the luminous matter is more clumped because of the following effect. There is a time for galaxy formation—why there is a critical time I shall come back to in a moment—and, of course, the largest amplitude fluctuations are the ones that would grow the fastest. Suppose that at this critical time, only matter that is three sigma deviations from the overall background has formed into proto-galaxies, formed into stable entities which will later become galaxies. Those clusters containing proto-galactic matter will tend to be highly concentrated in regions where there are upward fluctuations on large spatial scales.

Now, we have an intriguing qualitative picture of clumps of real and also clumps of "failed" galaxies. It would be marvelous to verify. We need good three-dimensional pictures of the distribution of galaxies. Well, for galaxies, of course, what you see in the sky is a two-dimensional projection. Fortunately, for distant galaxies we can also infer their distance by the fact that the universe is expanding, so that those further away from us are moving away from us faster than the nearer ones. So by combining red shift information and the position in the sky, we can get three-dimensional maps and begin to look for structures in three dimensions. I talked about galaxies, and we know their two-dimensional clumping, of course, but to predict something new, we have to look at something new, and here there is a marvelous idea of Mark Davis and a variety of other people. First of all, I should say this: it is crucial to have a lot of data in this game, and it is very important that one do automated red shift studies, process large numbers of galaxies at once, and get the real picture. Also, we can look at the failed galaxies, because although they do not light up themselves, they will absorb light, in particular from occluded quasi-stellar objects (which are the furthest objects we can see). So by studying the absorption of quasi-stellar light, you can look for failed galaxies: we have, first of all, to do the difficult measurements involved in finding them, then check that they are correlated among themselves and anticorrelated with the normal galaxies. So that is one major synthesis I see emerging. It is certainly nontrivial and unexpected, and we are very hopeful for the future. It is remarkable that initial conditions suggested by particle physicists are leading to very, very nontrival predictions for galaxy formation.

Let me also mention another thing, that in these studies, we are also interested in more detailed questions about galaxy formation, which I do not have time to go into right now—the question of how their morphology depends on their environment, do elliptical galaxies occur near other elliptical galaxies, and so forth, which tells us about how they were formed, whether there are large voids, how these structures look in three dimensions and so forth.

The general theme which I can extract from this, then, is that here and from other examples, we have developed over the course of years, and especially in the 20th century, methods of extraordinary power for analyzing matter—things like the Mössbauer effect, nuclear magnetic resonance, neutron activation, spectroscopy in all parts of the electromagnetic spectrum and so forth. But for the most part, except for the traditional disciplines of microscopy and telescopy, these disciplines have mostly been giving us crude averaged measurements; I mean, for instance, measurements of purity, measuring amounts of material in bodies as a whole. With the onset of the possibility of processing huge amounts of information, it becomes possible to do something more interesting, I think, on a large scale in many problems, and that is to trace the development of structures in space and time and to look at more subtle things. It is going to be important to go beyond that, to develop ways—I do not know if there are general ways; it may vary from problem to problem—but different ways of identifying structures that are not simple correlation functions, and to make use of all this data we are going to be collecting on development of structures in space and time. Some examples of imaging technologies, at various stages of development and sensitive to very different things, are: CAT scans, NMR imaging, and Josephson arrays. I think for the first time, there is a qualitative difference here, that we are developing new ways of decoding matter. It is a trite saying that our various tools expand the senses of man, but I think it is, for the first time, becoming true that these tools are becoming comparable in sophistication to the senses of man, but in different regions of the spectrum and sensitive to different things, and we shall have to develop the brain power, together with the sensory power, to make use of all this information. Now, we typically deal with these things by turning the output into pictures and exploiting our visual processing abilities. Eventually it might be too much for our brains, but we have computers now to help us out.

By the way, real sensory systems also provide highly nontrivial physical problems which we may learn from. For instance, the ear has a very, very impressive performance, being able to sense amplitude vibrations of the order of 10^{-9} cm, and actually reaches the quantum limits of sensitivity taking measurements over a millisecond which, when you think about it, is an extraordinary physical achievement. Bill Bialek, in particular, has developed very interesting theoretical models of how that can happen.

II. THEORETICAL EXPERIMENTS

Another development that I perceive having a great future—it has a great present—is the development of purely theoretical experiments. Steve Wolfram told us about one of those yesterday, and I think it is a very broad theme that runs through a variety of things, but as always I shall go from the specific to the general. A specific example of supreme importance is that we have every reason, since the mid-70's, to think that we have the correct microscopic theory of the strong interaction. We can write down the Langrangian for quarks' interactions with gluons, and this should tell us everything we need to know about this form of interaction: what physical particles come out, scattering amplitudes, nuclear energies and so forth. However, the things we can actually compute from this grandiose vision are relatively limited. In fact, as I look at my data book from 1980, there is literally nothing you can compute accurately about the strong interaction. In the new edition of the data book, they have data on scaling violations, and now there is an extra page of things that you can compute. It does work. We are pretty sure the theory is correct, and there are any number of qualitative indications that it is correct. But there are many significant qualitative and certainly quantitative problems that elude us, and it is very frustrating to have this theory and not be able to compute very many of the things which originally motivated us to formulate it. For instance, we would very much like to know whether you can form new forms of matter involving quarks, in unusual conditions of high pressure or density which you might find in neutron stars or heavy-ion collisions, or at high temperature which you might find in heavy-ion collisions or certainly in the early universe. Then there is the challenge of computing what is in the rest of this book: computing things in nuclear physics or justifying nuclear physics models, for example. To answer these questions requires doing integrals, but integrals on a grand scale, integrals by the hundreds of millions every second to do meaningful simulations. These things require the most powerful computers; new numerical methods and even new architectures of computers are being developed to handle this very problem, and solutions are in sight.

Now perhaps there is nothing qualitatively new here; of course, people have used whatever computational tools were available for a long time to solve the problems that they wanted to solve, but I think it has become qualitatively different. One can also do things in the theoretical experiment that you can not do in ordinary experiments. For instance, in a quark gluons theory, it is believed that most of the interesting dynamics, the underlying deep structure of what is going on, is due to some kind of fluctuations in the gluon field. Some people advertise instantons, Feynman has some other ideas about what the important configurations are, and so on. Gluons do not interact with any of our ordinary electromagnetic probes in a very direct way. It is very hard to get at those fluctuations by measurements. But on a computer you can ask whatever theoretical question you want to ask. So we find in these theoretical experiments once more the same problem of seeing patterns. Now that we have perfect flexibility—we can ask any questions we want—we have to be able to isolate what patterns are going to be pertinent. There is also the possibility of varying problems, in this case things like the number of quarks, or their masses, which you do not get to vary in the laboratory very easily.

There are several other major identifiable problems in a similar state as QCD. One on which progress could be exceedingly significant is the problem of translating one-dimensional information that we know about genes into some knowledge about the three-dimensional and even four-dimensional structures of biomolecules. There are now very efficient methods of reading off sequences so that we know the information that is necessary to construct these three-dimensional molecules, but very little progress—it is not fair to say very little progress—but the decisive progress has not been made on translating that one-dimensional information, which is the complete information, after all, into three-dimensional information that enables us to design drugs and to understand much better all kinds of processes in the cell.

T. PUCK: I do not understand; are you proposing to design drugs purely from a structural...

F. WILCZEK: I certainly do not know how to go about it, but I certainly think it would be very useful in designing drugs to know what the molecules you are thinking about look like; see whether they would fit with what you want to plug them into.

T. PUCK: Well, it would, and the fact of the matter is that there are very powerful new methods coming out in the design of drugs.

H. FRAUENFELDER: I think what you [Puck] are talking about is at a much lower level than this because you assume that much of the structure is already known—not you, but the people who design these drugs. It is a much simpler problem than here, where you really start from the one-dimensional information.

T. PUCK: That is what I presumed, but what I wanted to know is, is it the abstract problem that is important or actually securing drugs?

F. WILCZEK: Well, there is much more to it than drugs. The general problem is that one can read out in very efficient ways, very powerful ways, one-dimensional gene sequences, and we would like to know from that one-dimensional information what a three-dimensional protein would look like and how does it function in time?

Star formation is another very significant problem. It impacts, for instance, on the galaxy formation questions. You have to know how star formation works to really understand that. Turbulence is another example. A very small inroad on that problem has been made recently by Feigenbaum and others. The great bulk of the problem lies in the future.

III. AN EXAMPLE

Finally, I would like to talk about something that I have been thinking about recently, and I shall be brief. This is called High Tech Meets Low-Energy Neutrinos and includes some things I have been thinking about with Blas Cabrera and Lawrence Krauss. It illustrates another trend in science that I think is very significant. Several major problems in high-energy physics and astronomy involve detection of low-energy neutrinos, low-energy in this sense meaning from 0.1 to 10 MeV. Such energies characterize neutrinos and anti-neutrinos that originate from nuclear reactors, from antural radioactivity and also from the sun. You would like to know the masses and possibly mixing angles of different neutrinos, and the most sensitive way of probing that is oscillation experiments at accelerators, reactors and for the sun. The slower you get the neutrinos, the more oscillations there are to see. And solar emission: we would like to know that the sun is really turned on—a solar neutrino experiment has not seen anything, which calls that into doubt. In any case, we would like to verify directly models for the interior of the sun—see the spectrum of neutrinos that come out directly without interacting anywhere in between. The difficulty, of course, is that solar neutrinos have exceedingly low cross-sections: 10^{-45} cm^2.

Now I would like to go into an apparent digression about the properties of silicon. Of course, it is the miracle substance of high tech. It also has great use for us. At low temperatures, the specific heat of silicon goes to very, very low levels, because it is an insulator; it has no electronic specific heat leading to T^3 specific heat, and it has a very large Debye temperature because it is strongly bonded andlight. As a result, the quantitative result is extremely impressive. I was shocked to learn that, if you deposit 0.1 MeV of energy into a mole of silicon at very low temperatures, you get heating to 4 mK. When you combine this with the fact that it is now feasible, using dilution refrigerators, to cool large amounts of matter to the few mK range and also relatively simple to measure this kind of temperature change at low temperatures, the possibility of a new kind of neutrino detector comes on the horizon.

Now we come to the problem of dirt, literally. The other useful feature of silicon is that it is available in extremely high purity. Its impurity levels, I have learned by three weeks on the telephone, are carbon of 10^{-8} and oxygen at 10^{-10}, and all others, insofar as you can measure them, less than 10^{-12}. So natural radioactivity, which in any other substance would be a severe problem, background decay of the detector as you are looking at it, is not apparently a severe problem with silicon. It appears quite feasible to get to the levels we need for the most interesting experiments.

The final great advantage of silicon is that it is relatively cheap by the standards of solar neutrion experiments or Santa Fe Institutes. For the sun, you need about 10^3 kg, a ton of stuff, to get two events per day; this would cost about $200,000.

So, that is a specific thing which I find very exciting. What can we learn from it in general? We are trying to see what physics is going on at 10^{12-15} GeV (where Gell-Mann, Ramond and Slansky taughts us that neutrino masses are probably determined) by doing measurements of their interactions at 1 MeV, and trying to measure the interactions of the sun, which is 10^7 degress Kelvin, by working at 10 mK. We are doing things which involve particle, nuclear, astro, low-temperature and material physics in a very nontrivial way. I think this illustrates the lesson that the Institute is planning to be dedicated to: namely, that science is organized around problems, specific problems—not around pre-existing frameworks—and to speak the language of the problem, we do whatever is necessary.

Note added 6/26/87: On looking over the preceding text, its origin as a transcript of an informal talk is all to apparent; nevertheless, I have done only minimal editing. Perhaps the rough style is not inappropriate to the unfinished state of the problems discussed.

I would like to add two brief remarks on developments in the interim. The idea of automatic abstraction of patterns from an ensemble of examples has been at the heart of research on leanring in neural nets. Some particularly impressive demonstrations of this approach are due to Rumelhart, Sejnowski, and their collaborators.

The idea of bulk cryogenic detectors, our Example, is being vigorously pursued by over a dozen groups around the world. The motivations for these efforts include not only neutrino detection, but also x-ray astronomy and–most remarkably–the detection of the cosmological "dark matter."

STEPHEN WOLFRAM
Center for Complex Systems Research, University of Illinois, Champaign-Urbana

Complex Systems Theory[1]

Some approaches to the study of complex systems are outlined. They are encompassed by an emerging field of science concerned with the general analysis of complexity.

Throughout the natural and artificial world, one observes phenomena of great complexity. Yet research in physics and to some extent biology and other fields has shown that the basic components of many systems are quite simple. It is now a crucial problem for many areas of science to elucidate the mathematical mechanisms by which large numbers of such simple components, acting together, can produce behaviour of the great complexity observed. One hopes that it will be possible to formulate universal laws that describe such complexity.

The second law of thermodynamics is an example of a general principle that governs the overall behaviour of many systems. It implies that initial order is progressively degraded as a system evolves, so that in the end a state of maximal disorder and maximal entropy is reached. Many natural systems exhibit such behaviour. But there are also many systems that exhibit quite opposite behaviour,

---

[1] Based on a talk presented at a workshop on "A response to the challenge of emerging syntheses in science," held in Santa Fe, NM (October 6-7, 1984). Revised January, 1985.

transforming initial simplicity or disorder into great complexity. Many physical phenomena, among them dendritic crystal growth and fluid turbulence are of this kind. Biology provides the most extreme examples of such self-organization.

The approach that I have taken over the last couple of years is to study mathematical models that are as simple as possible in formulation, yet which appear to capture the essential features of complexity generation. My hope is that laws found to govern these particular systems will be sufficiently general to be applicable to a wide range of actual natural systems.

The systems that I have studied are known as cellular automata. In the simplest case, a cellular automaton consists of a line of sites. Each site carries a value 0 or 1. The configurations of the system are thus sequences of zeroes and ones. They evolve in a series of time steps. At each step, the value of each site is updated according to a specific rule. The rule depends on the value of a site, and the values of, say, its two nearest neighbors. So, for example, the rule might be that the new site value is given by the sum of the old value of the site and its nearest neighbours, reduced modulo two (i.e., the remainder after division of the sum by two).

Even though the construction of cellular automata is very simple, their behaviour can be very complicated. And, as a consequence, their analysis can be correspondingly difficult. In fact, there are reasons of principle to expect that there are no general methods that can universally be applied.

The first step in studying cellular automata is to simulate them, and see explicitly how they behave. Figure 1 shows some examples of cellular automata evolving from simple seeds. In each picture, the cellular automaton starts on the top line from an initial state in which all the sites have value zero, except for one site in the middle, which has value one. Then successive lines down the page are calculated from the lines above by applying the cellular automaton rule at each site. Figure 1(a) shows one kind of pattern that can be generated by this procedure. Even though the rule is very simple (it can be stated in just one sentence, or a simple formula),

FIGURE 1 Patterns generated by evolution according to simple one-dimensional cellular automaton rules from simple initial conditions.

FIGURE 2   Snowflake growth simulation with a two-dimensional cellular automaton (courtesy of Norman H. Packard).

and the initial seed is likewise simple, the pattern produced is quite complicated. Nevertheless, it exhibits very definite regularities. In particular, it is self-similar or fractal, in the sense that parts of it, when magnified, are similar to the whole.

Figure 2 illustrates the application of a cellular automaton like the one in figure 1(a) to the study of a natural phenomenon: the growth of dendritic crystals, such as snowflakes (as investigated by Norman Packard). The cellular automaton of figure 1(a) is generalized to be on a planar hexagonal grid, rather than a line. Then a cellular automaton rule is devised to reproduce the microscopic properties of solidification. A set of partial differential equations provide a rather complete model for solidification. But to study the overall patterns of growth produced, one can use a model that includes only some specific features of the microscopic dynamics. The most significant feature is that a planar interface is unstable, and produces protrusions with some characteristic length scale. The sizes of the sites in the cellular automaton correspond to this length scale. The sizes of the sites in the cellular automaton correspond to this length scale, and the rules that govern their evolution incorporate the instability. With this simple caricature of the microscopic laws, one obtains patterns apparently very similar to those seen in actual snowflakes. It remains to carry out an actual experiment to find out whether the model indeed reproduces all the details of snowflakes.

Figure 1(b) shows a further example of a pattern generated by cellular automaton evolution from simple initial seeds. It illustrates a remarkable phenomenon: even though the seed and the cellular automaton rules are very simple, the pattern produced is very complicated. The specification of the seed and cellular automaton rule requires little information. But the pattern produced shows few simplifying features, and looks as if it could only be described by giving a large amount of information, explicitly specifying its intricate structure.

Figure 1 is a rather concrete example of the fact that simple rules can lead to very complicated behaviour. This fact has consequences for models and methodologies in many areas of science. I suspect that the complexity observed in physical processes such as turbulent fluid flow is of much the same mathematical character as the complexity of the pattern in figure 1(b).

The phenomenon of figure 1 also has consequences for biology. It implies that complicated patterns of growth or pigmentation can arise from rather simple basic processes. In practice, however, more complicated processes may often be involved. In physics, it is a fair principle that the simplest model for any particular phenomenon is usually the right one. But in biology, accidents of history often invalidate this principle. It is only the improbability that very complicated arrangements have been reached by biological evolution which makes a criterion of simplicity at all relevant. And, in fact, it may no more be possible to understand the construction of a biological organism than a computer program: each is arranged to work, but a multitude of arbitrary choices is made in its construction.

The method of investigation exemplified by figures 1 and 2 is what may be called "experimental mathematics." Mathematical rules are formulated, and then their consequences are observed. Such experiments have only recently become feasible, through the advent of interactive computing. They have made a new approach to science possible.

Through computers, many complex systems are for the first time becoming amenable to scientific investigation. The revolution associated with the introduction of computers in science may well be a fundamental as, say, the revolution in biology associated with the introduction of the telescope. But the revolution is just beginning. And most of the very easy questions have yet to be answered, or even asked. Like many other aspects of computing, the analysis of complex systems by computer is an area where so little is known that there is no formal training that is of much advantage. The field is in the exciting stage that anyone, whether a certified scientist or not, can potentially contribute.

Based on my observations from computer experiments such as those of figure 1, I have started to formulate a mathematical theory of cellular automata. I have had to use ideas and methods from many different fields. The two most fruitful so far are dynamical systems theory and the theory of computation.

Dynamical systems theory was developed to describe the global properties of solutions to differential equations. Cellular automata can be thought of as discrete idealizations of partial differential equations, and studied using dynamical systems theory. The basic method is to consider the evolution of cellular automata from all its possible initial states, not just, say, those consisting of a simple seed, as in figure 1. Figure 3 shows examples of patterns produced by the evolution of cellular automata with typical initial states, in which the value of each site is chosen at random. Even though the initial states are disordered, the systems organizing itself through its dynamical evolution, spontaneously generating complicated patterns. Four basic classes of behaviour are found, illustrated by the four parts of figure 3.

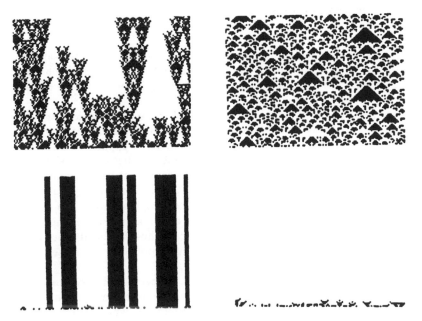

FIGURE 3 Four classes of behaviour found in evolution of one-dimensional cellular automata from disordered initial states.

The first three are analogous to the fixed points, limit cycles and strange attractors found in differential equations and other dynamical systems. They can be studied using quantities from dynamical systems theory such as entropy (which measures the information content of the patterns), and Lyapunov exponents (which measure the instability, or rate of information propagation).

Cellular automata can not only be simulated by computers: they can also be considered as computer in their own right, processing the information corresponding to their configurations. The initial state for a cellular automaton is a sequence of digits, say, ones and zeroes. It is directly analogous to the sequence of digits that appears in the memory of a standard digital electronic computer. In both cases, the sequences of digits are then processed according to some definite rules: in the first case the cellular automaton rules, and in the second cases the instructions of the computer's central processing unit. Finally, some new sequence of digits is produced that can be considered as the result or output of the computation.

Different cellular automata carry out computations with different levels of complexity. Some cellular automata, of which figure 3(d) is probably an example, are capable of computations as sophisticated as any standard digital computer. They can act as universal computers, capable of carrying out any finite computation, or of performing arbitrary information processing. The propagating structures in figure 3(d) are like signals, interacting according to particular logical rules.

If cellular automata such as the one in figure 3(d) can act as universal computers, then they are in a sense capable of the most complicated conceivable behaviour. Even though their basic structure is simple, their overall behaviour can be as complex as in any system.

This complexity implies limitations of principle on analyses which can be made of such systems. One way to find out how a system behaves in particular circumstances is always to simulate each step in its evolution explicitly. One may ask whether there can be a better way. Any procedure for predicting the behaviour of a system can be considered as an algorithm, to be carried out using a computer. For the prediction to be effective, it must short cut the volution of the system itself. To do this, it must perform a computation that is more sophisticated than the system itself is capable of. But, if the system itself can act as a universal computer, then this is impossible. The behaviour of the system can, thus, be found effectively only by explicit simulation. No computational short cut is possible. The system must be considered "computationally irreducible."

Theoretical physics has conventionally been concerned with systems that are computationally reducible and amenable, for example, to exact solution by analytical methods. But I suspect that many of the systems for which no exact solutions are now known are, in fact, computationally irreducible. As a consequence, at least some aspects of their behaviour, quite possibly including many of the interesting ones, can be worked out only through explicit simulation or observation. Many asymptotic questions about their infinite time behaviour, thus, cannot be answered by any finite computations, and are thus formally undecidable.

In biology, computational irreducibility is probably even more generic than in physics and, as a result, it may be even more difficult to apply conventional theoretical methods in biology than in physics. The development of an organism from its genetic code may well be a computational irreducible process. Effectively the only way to find out the overall characteristics of the organism may be to grow it explicitly. This would make large-scale computer-aided design of biological organisms, or "biological engineering," effectively impossible: only explicit search methods analogous to Darwinian evolution could be used.

Complex systems theory is a new and rapidly developing field. Much remains to be done. The ideas and principles that have already been proposed must be studied in a multitude of actual examples. And new principles must be sought.

Complex systems theory cuts across the boundaries between conventional scientific disciplines. It makes use of ideas, methods and examples from many disparate fields. And its results should be widely applicable to a great variety of scientific and engineering problems.

Complex systems theory is now gaining momentum, and is beginning to develop into a scientific discipline in its own right. I suspect that the sociology of this process is crucial to the future vitality and success of the field. Several previous initiatives in the direction of complex systems theory made in the past have failed to develop their potential for largely sociological reasons. One example is cybernetics, in which the detailed mathematical results of control theory came to dominate the field, obscuring the original, more general goals. One of the disappointments in complex systems theory so far is that the approaches and content of most of the papers that appear, reflect rather closely the training and background of their authors. Only time will ultimately tell the fate of complex systems theory. But as of now the future looks bright.

REFERENCES

S. Wolfram, "Computer Software in Science and Mathematics," *Scientific American* (September, 1984).

S. Wolfram, "Cellular Automata as Models of Complexity," *Nature* **311** (1984), 419–424.

S. Wolfram, "Undecidability and Intractability in Theoretical Physics," *Physical Review Letters* **54** (1985), 735.

S. Wolfram, "Twenty Problems in Theory of Cellular Automata," to be published in *Physica Scripta*.

S. Wolfram, "Origins of Randomness in Physical Systems," submitted to *Physical Review Letters*.

FELIX E. BROWDER
Department of Mathematics, University of Chicago

Mathematics and the Sciences

One of the most striking features of the development of fundamental theory in the sciences during the past decade has been the convergence of its focal interests with major themes in mathematical research. Mathematical concepts and tools which have arisen in an apparently autonomous way in relatively recent research have turned out to be important as major components of the description of nature. At the same time, this use of novel mathematical tools in the sciences has reacted back upon the development of mathematical subject matter having no obvious connection with the scientific subject matter to yield new and surprising mathematical consequences. It is this theme of strong reciprocal interaction which I propose to present in the present discussion. We must ask why this has been so, whether this kind of interaction is a major trend that will continue in a serious way into the foreseeable future and, if so, what the consequences will be for the future development of mathematics and the sciences.

SECTION I: MATHEMATICS AND THE NATURAL SCIENCES

Let us begin our analysis by examining the different ways in which novel and relatively sophisticated mathematical tools have been applied in recent scientific developments. We may classify them into five relatively broad modes of attack.

1. THE USE OF SOPHISTICATED MATHEMATICAL CONCEPTS IN THE
FORMULATION OF NEW BASIC PHYSICAL THEORIES ON THE MOST
FUNDAMENTAL LEVEL. At the present moment, this takes the form of the
superstring theory which has as its objective the total unification of all the basic
physical forces and interactions: electromagnetic, weak, strong, and gravitational.
This new phase of physical theory which is the culmination of the earlier development of gauge field theories and of theories of supersymmetry exhibits the use
of a wide variety of relatively new mathematical tools developed in the past two
decades such as Kac-Moody algebras and their representations, the existence of
Einstein metrics on compact Kahlerian manifolds satisfying simple topological re-
strictions, and representations of exceptional Lie groups. The body of techniques
and mathematical arguments embodied here includes the theory of Lie groups and
algebras, their generalizations, and their representation theory, differential geometry in its modern global form in terms of vector bundles, the study of the existence
of solutions of on manifolds of highly nonlinear, partial differential equations, differential and algebraic topology, and the whole melange of analysis, algebra and
geometry on manifolds which has been called *global analysis*. The implementation
of this program involves still other major directions of mathematical research, most
particularly problems in algebraic geometry.

A similar pattern of the use of sophisticated mathematical tools in the development of fundamental physical theories appeared earlier in the context of the study
of *instantons* in gauge field theories, and of the study of singularities in the equations of general relativity in connection with *black holes*. What must be strongly
emphasized in all these cases is that the role assumed by sophisticated mathematics
was not the result of a willful act by either physicists or mathematicians, but of
the intrinsic necessities of the development of the physical theory. Physicists, no
matter how sophisticated mathematically they may be, are not free ad libitum to
chose the mathematical tools they wish to use. Certainly the mathematicians have
no power to prescribe such uses to the physicists. We are very far from the decades
after the Second World War when it was a commonplace among physicists that all
the mathematics they would ever need had been completely worked out (at least
as far as the involvement of research mathematicians was concerned) by the time
of the First World War. It is the radical transformation of fundamental physics in
the past decades that has caused the disappearance of this commonplace, and not
any basic transformation in the sociology of the relations between physicists and
mathematicians.

2. A FOCAL INTEREST ON THE COMPLEX MATHEMATICAL CONSE-
QUENCES OF SIMPLE PHYSICAL LAWS. One sees major examples of this trend
in the modelling of turbulence in terms of bifurcation, of the asymptotic properties

of differential equations and iteration of simple nonlinear transformations (Hopf bifurcation, the Lorenz equation, strange attractors, and Feigenbaum cascades). Very simple causal mechanics can be shown to lead to disorderly regimes (*chaos*), but in relatively simple and classifiable forms. An historically earlier example of an attack on turbulence in the 1930's to 1950's used models in terms of stochastic processes where disorder was directly injected into the premises of the theory. Another current example is the use of fractal models (self-similarity under changes of scale, fractional Hausdorff dimensions) to describe complex phenomena in the study of materials.

3. MATHEMATICAL MODELS OF PATTERN FORMATION AND SYMMETRY BREAKING AS PARADIGMS FOR STRUCTURED SYSTEMS DEVELOPING OUT OF APPARENTLY UNSTRUCTURED REGIMES. We might think of this mode of attack (which goes back to a paper of Turing in 1952) as the converse of (2). Stable structures are seen to arise from mathematical models of differential equations or stochastic games of an apparently structureless nature in the presence of noise and possible disorder. The most striking paradigm is the oscillating chemical reaction of the Belousov-Zhabotinskii type. The objective here is to eventually model phenomena in such areas as developmental biology and brain function.

4. SOLITON THEORIES, INVOLVING THE EXISTENCE IN NONLINEAR DIFFERENTIAL EQUATIONS OF STABLE STRUCTURES (SOLITONS) ARISING FROM COMPLETE INTEGRABILITY. The now classical paradigm is the Korteweg-De Vries equation of shallow wave theory, rediscovered by M. Kruskal and his collaborators in the late 1950's after an earlier partial rediscovery in computer experiments by Fermi-Pasta-Ulam. New models of a similar kind have been found and extensively analysed as a possible way of describing a broad range of physical and engineering phenomena.

5. THE FELT NEED TO DEVELOP A USABLE AND FRUITFUL MATHEMATICAL THEORY OF COMPLEX SYSTEMS WHOSE ELEMENTS MIGHT WELL BUT WHOSE COMPLEXITY ARISES FROM THE INTERACTION OF THESE ELEMENTS, WHETHER LINEAR OR NONLINEAR, LOCAL OR GLOBAL. It is abundantly clear that every mode of analysis in science or in practice will eventually get to the stage where this theme is dominant, and in most cases sooner rather than later. Topics (2), (3), and (4) are simply some of the currently active sub-themes of the overall theme of complexity in the presence sense.

Having presented this very summary description of major thematic components in present-day scientific investigations in which mathematical tools of a relatively sophisticated kind are being applied, we may ask whether this is really a new situation. A careful answer to this question demands another kind of analysis with a historical and philosophical focus which we present in the latter part of the present paper.

SECTION II. MATHEMATICS AND THE COMPUTER

The observant reader will already be aware that in the description presented in Section I of the mathematical component of important themes in contemporary scientific research, no explicit mention was made of the high-speed digital computer, one of the most conspicuous objects of our age. In the context of the present kind of discussion, this might seem to many like a performance of Hamlet without the Noble Dane. Yet we must, in fact, segregate the discussion of the computer and its interrelation with the development of contemporary mathematics, both because of its important and distinctive role and because of the prevalence and intensity of myths in this domain which confuse and disable realistic assessment of the situation.

We are all very conscious of the decisive role of the high-speed digital computer as one of the decisive facts of the present epoch and as far as we can see into the foreseeable future. We all know of the tremendous impact it has already had, which promises to be even more accentuated in the future, on the structure of all processes in the world industrial society which depend on calculation, communication and control. In practice, this excludes very few domains of human existence in modern society, whether technological, economic, social, political, or military. There is no reason to believe that the sciences or mathematics can be immune from this kind of impact; indeed, the scope and nature of scientific and mathematical instrumentation and practice in our society have already been radically changed by the existence of high-speed digital computation and its continual decrease in cost during recent decades. I have deliberately used the unusual phrase *mathematical instrumentation* to point up the fact which is radically new that such a phenomenon now exists and is an important component of our present-day situation.

At the same time, while we are all conscious of the importance of the digital computer (sometimes to the point of hysteria) and indeed are inundated with advertising hype from the most diverse quarters about all the wonders that supercomputers will do for us, many are much less conscious of what is ultimately an even more important fact: the computer is as much a *problem* as it is a *tool*. We must understand the nature and limitations of this most powerful of all human tools. It is important to know *what cannot be computed* and the dangers of *what can be mis-computed*.

This can be seen most plainly and with the least equivocation in the context of mathematical and scientific practice. Perhaps the most significant use of the computer in this context is as an experimental tool, sometimes even displacing the laboratory experiment altogether. One translates a scientific or mathematical problem into a simpler mathematical model, and then uses the computational power of the computer to study particular cases of the general model. This has turned out to be a very useful approach on many occasions, particularly when the conditions for experiment in the usual sense or of precise calculation become impossibly difficult. The mystique of such practices has grown to such an extent that some speak of replacing Nature, an analog computer, by a newer and better model of a digitalized nature.

The drawbacks and dangers of such practices, without a background of thorough critical analysis, are equally clear. We must ask about the adequacy of the model, about the accuracy (not to say the meaningfulness) of the computational process, and, last but not least, about the representative character of the particular cases which one computes. Without serious cross-checks on all these factors, we are left with still another case of the zeroth law of the computer: *garbage in, garbage out*. This is particularly the case because of one major circumstance in very serious scientific and mathematical problems: they cannot be solved by computation as they stand. One replaces them by manageable problems, and the validity of the replacement is precisely the crucial question. It is the importance of this question that has led to ironical comments on the adjective *scientific* in the currently fashionable emphasis on programs for scientific computation on supercomputers.

A critical approach to such questions is by no means equivalent to any sort of advocacy of neglecting the computer as a tool in science and mathematics, not to speak of its other and even more important domains of application in society at large. It does point up a sometimes neglected fact: the computer is a difficult tool, and its use must be studied and refined. Computers are *brute force* instruments; their effective use depends vitally on human insight and ingenuity. Computers yield no insights by themselves, and their effective use depends upon the skill and insight of those who program them.

The thrust of these remarks is to put forward in a sharp way, the importance of the intellectual arts and insights which are and can be connected with the digital computer and its uses. These intellectual arts (organized under various labels, most commonly computer science) have a very vital relation to the mathematical enterprise. They constitute a specialized and different way of applying classical mathematical ideas and techniques with radically new purposes in mind. Their vitality, both intellectual and practical, depends in a very essential way upon a continuing contact with the central body of mathematical activity.

There is an interesting and slightly ironical aspect to this relationship of computer science with the central body of mathematics which has occasioned a great deal of discussion in recent years. Since the Scientific Revolution of the 17th century which laid down the original outline of the viewpoint of modern physical science, there has been a tendency among mathematicians and physical scientists to see a dichotomy within mathematics between two kinds of mathematics, that kind which is *applicable* to the uses of modelling and calculation in physical situations and another kind which is *not applicable*. The rules for this break-up have changed over the years, as we emphasized in Section I, with an ever-increasing diversity of mathematical themes and theories falling into the first class. Even so, the stereotype tends to persist, and some areas of active mathematical research like algebraic number theory (and the related area of algebraic geometry of characteristic p) or mathematical logic tend to be relegated to the second class. Yet it is precisely these areas, grouped together with various forms of combinatorics under the general label of discrete mathematics, which have turned out to be of the most vital significance in major areas of advance in computer science. The basic theoretical framework of computer science and the development of the study of complexity of computation rest upon the foundation of mathematical logic. The development of algorithms depends essentially upon combinatorics, number theory, and most recently on probabilistic models of a combinatorial type. The very practical area of coding and of cryptology, computer encryption and deciphrement, is vitally dependent upon sharp results in number theory and algebraic number theory.

Since computer science is a new and extremely vital branch of the sciences, we cannot wave away the scientific relevance of these branches of mathematics. Some devotees of this new wave of mathematical relevance have gone even further. Programs of mathematical instruction on the college level have usually begun with differential and integral calculus in the last fifty years, because the differential and integral calculus is the mathematical language and elementary underpinning of classical physics. The proponents of *discrete mathematics* suggest today that calculus should be replaced by a new course on this level, centering on combinatorics and number theory. New courses in so-called applied mathematics are being organized in undergraduate colleges throughout the country based upon a combination of discrete mathematics, the basics of computer programming, and a sprinkling of elementary statistics. It has yet to be proved that such courses can be taught as an adequate substitute for the more familiar basic curriculum, but the existence of this movement can be regarded as another symptom of the changing relations of mathematics to its potential domains of application and relevance.

As the tone of my last remarks would indicate, I regard the effort to produce a programmatic dichotomy between the discrete and the continuous to be a snare and a delusion, along with other systematic efforts to oppose the natural sciences to the "artificial" sciences. Human art and artifice are part of all the sciences, as is the confrontation with the objective realities beyond human will and control that we personify under the figure of Nature. Indeed, I should like to make the case that the computer science in its necessary advance, seen today under such perspectives as parallel processing, artificial intelligence and expert systems, and the whole family of problems subsumed under the label of computer systems and structure, is another subclass of the more general perspective that we described in Section I under the label *complexity of organization*.

SECTION III. THE CORE OF MATHEMATICS

There is a danger and an illusion in any form of discussion of the role of mathematics that emphasizes as I have done the active participation of new mathematical concepts and tools in the development of other scientific disciplines. Despite the strong emphasis on the *new*, it is far too easy to use such a description as a prescription that the appropriate role of mathematicians as such in the future is simply to facilitate the interactions that I have described. In my view, such a prescription would be a recipe for a massive *failure*, not only from the point of view of the development of mathematics itself, but from the point of view of the sciences. Prescriptions of this kind are based upon the unconscious principle that *creativity* and *newness* in conceptual advance are always a matter of the past (or at best of the relatively recent past). The *autonomy* of mathematical research, in the sense of its freedom from any strong dependence upon the current processes of research in other disciplines and upon their rhythym of activity, has been one of the principal components of its creativity. This has been the case through the whole lengthy history of mathematics going back to the Greeks. One of the obvious, common sense reasons why such autonomous mathematical creativity is important for the sciences is that, when the advance of scientific understanding needs mathematical concepts, theories, or methods of calculation and argument, it is very important and sometimes essential that they should already have been developed in a reasonably usable form. There is a sort of idle tribal vanity (derived from an even sillier form of personal vanity) that one group of scientific practitioners, the theorectical physicists, for example, could easily do the work of another group, the mathematicians, for example, better than the latter. The validity of such assumptions might be debated (if one had nothing better to do), but not the validity of another crucial point: whoever does the job is working as a mathematician and has to face the difficulties of solving mathematical problems. Once the problems are solved, the solutions can be digested and turned to new uses in other contexts. Yet the new mathematics involved (concepts, solutions, theorems, algorithms, proofs, and calculations), if it is genuinely new, must be created by someone, and whoever does the job is a *mathematician* by the definition of the latter term. The task of the practitioner of another scientific discipline with respect to mathematics is to use it to understand and analyse the subject matter of that discipline, to *see through* the mathematics to the structure of his own subject matter. From the fact that the mathematics from the latter point of view *ought to be transparent*, one cannot draw the false (though occasionally fashionable) conclusion that the mathematics does not exist and needs no process of development in its own right.

It may seem like a paradox to some that I should introduce this strong affirmation of the essential autonomy of mathematics into a paper devoted to the central theme of the interaction of mathematics and the sciences. This paradox is superficial. Any affirmation of interaction is only significant if the two sides of the interaction have a full-fledged separate existence and meaningfulness. In particular, we must affirm a central autonomous core of meaning in the mathematical enterprise if our thesis of strong interaction is to have its full significance.

What is this core meaning? I shall give a number of related answers in the form of programmatic definitions of *mathematics*. Each of these definitions points to important characteristics of mathematical practice, and each program leads to a slightly different perspective on that practice. It would take me too far afield in the present discussion to describe the interrelation of these perspectives and the tension between them. Suffice it to say that I am among those who believe in an essential unity of mathematics, though rejecting some of the dogmatic and over-simplified programs for achieving that unity by putting mathematics in a Procrustean bed and cutting off some of its limbs.

1. Mathematics is the science of significant forms of order and relation.
2. Mathematics is the science of the structure of possible worlds.
3. Mathematics is the science of infinity.
4. Mathematics is the science of the structure of complex systems.
5. Mathematics is the study of the modelling of reality in symbolic form.

Each of these definitions taken by itself is a deep truth in the sense of Niels Bohr; its negation is also a deep truth. Taken jointly, they give us a reasonable general perspective on the broad range of mathematics since the Renaissance. (Definitions 1 and 2 are due to Descartes and Leibniz, combined under the term *mathesis*, while Definition 3 which was originated by Leibniz was revived in modern times by Poincaré and Weyl.)

As I remarked earlier, mathematical research in its autonomous forms is an enterprise of great vitality in the present-day world (though somewhat invisible to most outsiders). As I stressed in the Introduction, despite its fundamental autonomy, the enterprise of front-line mathematical research has had a very strong interaction in the last two decades with various forms of advance in the sciences. For the purposes of the present discussion, in order to go beyond what was said in Sections 1 and 2 above, I present two kinds of evidence.

The first kind of evidence consists of taking a conventional breakdown of the principal active branches of contemporary mathematical research and inquiring in general terms whether these branches have interactions of the type described with the sciences. In the table of organization for the next International Congress of Mathematicians (to be held in Berkeley, California in the Summer of 1986), we have such a breakdown in the division of the Congress into 19 sections, namely

1. Mathematical logic and foundations
2. Algebra
3. Number theory
4. Geometry
5. Topology
6. Algebraic geometry
7. Complex analysis
8. Lie groups and representations
9. Real and functional analysis
10. Probability and mathematical statistics
11. Partial and differential equations
12. Ordinary differential equations and dynamical systems
13. Mathematical physics
14. Numberical methods and computing
15. Discrete mathematics and combinatorics
16. Mathematical aspects of computer science
17. Applications of mathematics to non-physical sciences
18. History of mathematics
19. Teaching of mathematics

Of these sections, (10), (13), (14), (16), and (17) by their definition relate directly to the sciences or to technology, while (18) and (19), of course, are not fields of mathematical research as such. A detailed analysis would reveal that the twelve other areas all relate in a strongly significant way either to the physical sciences or to computer science (or possibly to both). Thus, (1), (2), (3), and (15) have very strong interactions with the computer science side while the remainder, (4), (5), (6), (7), (8), (9), (11), and (12), bear as strongly upon the physical sciences.

The validity of this kind of analysis is best attested by those who fail to sympathize with it. As one such witness, I may cite the French mathematician Jean Dieudonné (one of the retired elder statemen of the Bourbaki group) who in a recent book surveying mathematical research in recent times under the title "Panarome des Mathematiques Pures. Le Choix Bourbachique," in each section gives a rather patronizing short squib (usually of a few lines) under the title "Rapports avec les Sciences de la Nature." In the first edition in French (1977), he was happy to announce of his favorite subject, algebraic geometry, that it has no relations with the natural sciences for the moment. That moment must soon have passed, or perhaps a better-informed reader had gotten in touch with him, for in the English edition (1982), the exemption is lifted and he remarks that algebraic geometry has interesting applications, both in the study of the Yang-Mills equations and in the theory of the Korteqeg-De Vries equation. (Of earlier examples, he says nothing.)

This last example, the study of the soliton theory of the Korteweg-De Vries equation in the periodic case, is actually an important illustration of the reverse process. The applications of algebraic geometry and complex analysis to the study of the Kortweg-De Vries equation under periodic boundary conditions not only contributed to the understanding of the physical model involved, but reacted back on the disciplines involved. New ideas and methods in both mathematical disciplines arose from this interaction, resulting in the solution of classical problems in algebraic geometry and function theory. In an even more striking case, it was observed by the young Oxford mathematician Simon Donaldson that, if one combined the mathematical techniques developed for the study of the mathematical theory of gauge fields by Schoen and Uhlenbeck with the penetrating geometrical attack upon the structure of four-dimensional manifolds of Michael Freedman, one could obtain a new and totally surprising geometrical result in four dimensions. The result in question asserts that unlike Euclidean spaces in every other dimension, four-dimensional Euclidean space possesses two systems of coordinates which are fundamentally different from one another.

These two cases illustrate a possibility turned into a current reality, that the strong mathematical attack upon mathematical problems raised in the context of development of research in the natural or even the social sciences can provide the occasion and stimulus for major conceptual advances in mathematics itself.

SECTION IV: PERSPECTIVES AND INSTITUTIONS

To close this essay, let me turn to the questions I posed at the beginning concerning the future relations of mathematics and the sciences, and try to relate these to the institutional context within which the various disciplines are pursued. In answering such questions, we may recall another well-known saying of Niels Bohr: prediction is difficult, especially of the future. Attempts to predict the future are indeed hypotheses about the past and present. I shall formulate such a hypothesis which we might check for coherence and accuracy against the past and present, and try to gauge its consequences for the future.

Let me begin with the distant past, with the beginning of the sciences in the civilization of the ancient Greeks. It was there that the concept of science as a self-conscious structuring of objective lawful knowledge of the world (or more strictly of the hidden processes of the world) first arose, and it was from the ancient Greeks that modern Western civilization inherited this concept as a distinctive heritage. Though the Greeks investigated the full range of their experience, their achievement in creating scientific knowledge that we continue to recognize as such was primarily in the mathematical sciences, in mathematics itself and in such highly mathematical disciplines as mathematical planetary astronomy, musical theory, and the mathematical treatment of statics. The Greeks created a highly perfected form of sophisticated mathematical theory treating of whole number, geometry, ratio, and geometrical measure. In this theory, they perfected, as well, a fully mature concept of mathematical argument, of logical deduction. On the basis of these achievements, Plato might argue in his celebrated dialogue *Timaeus* for a mathematical *myth* of the cosmos and its formation on the basis of geometrical elements, while Aristotle could formulate the logical principles of deduction while rejecting the possibility of mathematical laws for the phenomena of terrestrial physics.

It is very fashionable to talk of scientific revolutions. On the most fundamental level, there has been only one scientific revolution, that of the 17th century in which modern science was formed. The concept of science which this century produced gave a description of the cosmos, the physical universe, in terms of the geometry of space and of numerical relations, a description which applied to both the skies and the earth. It saw this cosmos as a realm of objective lawful relations, devoid of human agency or affect. Reality was separated after Descartes into two completely distinct parts, the physical universe and a separate world of human consciousness and spirit. In this framework, it made total sense for human consciousness to try to determine the secrets of natural processes not by passive observation, but by transforming nature by experiment, putting it to the test of torture following the best judicial precepts of the age.

There was a mathematical counterpart of the new physical science, which served both as its precursor and principal tool. This was the mathematics of the new algebra and of the analytic movement of Vieta and Descartes, a mathematics which substituted calculation and manipulation of symbolic expressions for the deductive sophistication of the Greeks. It substituted the analysis of complex phenomena into simple elements for the Greek synthetic transformation of simple axioms and principles into the complexities of deduced conclusions. In the 17th century, this new mathematics had two overwhelming triumphs: the creation of an analytical geometry through which the geometric structure of space could be transformed by coordinatization into the subject matter of algebraic analysis, and the invention of the great analytic engine of the differential and integral calculus by which the sophisticated and difficult arguments by exhaustion of Eudoxus and Archimedes for handling infinite processes were replaced by much simpler and more manageable algebraic formulae or *calculi*. This was the tool with which Newton built his great mathematical world-machine, the central paradigm for the scientific world pictures of all succeeding ages.

There are essentially two forms in which objective human knowledge can be formulated, in words and in mathematical forms. Aristotle opted for the first and created a systematic description of the world in which the subject-predicate form of the sentence was transformed into the pattern of the individual object or substance possessing a certain quality. From the 17th century on, modern science has rejected this form of description and replaced it by descriptions in various mathematical forms. These forms have altered as the stock of mathematical forms has increased and become richer and more sophisticated. The original forms were geometric, in the style of the Greeks. In the Renaissance, a new and more flexible concept of number, the "real" number in the present-day sense, came into being as the common measure of lengths, areas, volumes, masses, etc. without the precise distinction between these measures in terms of geometrical form to which the Greeks had held for their own very good intellectual reasons. In the ensuing development of algebra, new kinds of "number" appeared as the solutions of algebraic equations. Since they were not numbers in the old sense, some were called "imaginary" and mixtures of the two types were called "complex." It was not until the end of the 18th century that these "complex" numbers were fully naturalized as members of the common sense mathematical realm by being identified in a simple way with the points of a Euclidean plane, the complex plane.

Since the 17th century, the enterprise of the scientific description of nature has continued to develop within this mathematical medium which was dimly foreshadowed by Plato's mathematical world-myth. As new scientific disciplines developed, they too entered the same framework of numerical relationship, geometrical form in space, and formulation of basic principles in mathematically expressed laws. As Kant put it in a well-known aphorism: "In every special doctrine of nature only so much science proper can be found as there is mathematics in it." In the nearly four centuries that have elapsed since Galileo began the 17th century scientific revolution, the curious relationship of autonomy and mutual dependence between the natural sciences and mathematics has taken ever more complex and sophisticated forms. The mathematical medium in which the various sciences live has continued to develop and take on new shapes. In the early 19th century, the intuitive concept of symmetry applied to the study of the roots of algebraic equations gave rise to the concept of group, which passing through the medium of its application to geometry and differential equations became in the 20th century the most essential building block of the fundamental description of the physical universe. The concept of space, enriched by the insights of Gauss and Riemann, gave rise to the richer geometrical concepts of Riemannian manifold and of curvature, through which the theory of general relativity of Einstein formulated. Through the analysis of integral equations and differential equations in the early 20th century, the concept of an infinite dimensional vector space was born, and the especially rich concept of a Hilbert space and Hermitian operators on a Hilbert space with their spectral theory to serve as the eventual underpinning of the formal structure of quantum mechanics. These are three of the most important examples of a very broad phenomenon.

New concepts and theories arise in mathematical research through the pressure of the need to solve existing problems and to create intellectual tools through which already existing mathematical theories and structures can be analysed and understood. Once the new concepts and theories become established, they themselves become the focus of intensive investigation of their own structure. The new is achieved through the medium of mathematical constructions, by which the new concepts and structures are given definite form. Though the imaginative process is free in some ultimate sense, its result once produced becomes a new objective realm of relationship of a determinate character. It is investigated by classical tools like deduction and calculation to establish its properties. This leads to new technical problems, which may eventually demand new concepts and constructions for their solution. The jump of insight and imagination that leads to new mathematical breakthroughs belies the stereotype of mathematical activity as an automatic machine-like process of mechanical application of formal rules. In its most extreme form, this emphasis upon the mathematical imagination is expressed in a classical anecdote about the great German mathematician David Hilbert of the earlier part of the 20th century who was celebrated both for his mathematical insight and his provocative modes of self-expression. He was asked about one of his former pupils who had disappeared from Gottingen. He replied that Herr X did not have enough imagination to become a good mathematician; he had become a poet instead.

Mathematical research, as a whole, balances the *radical* process of generation of new concepts and theories with the *conservative* tendency to maintain in existence all those domains, problems, and conceptual themes that once become established as foci of significant mathematical research. The balance between these two opposing tendencies gives rise to the striking fact that at the same moment, one can find active research programs of apparently equal vitality bearing on themes, one of which is two thousand years old, while the other is only a few decades old. Yet the two-thousand-year-old problem might well be solved with tools and concepts of relatively recent vintage. Thus, the problems of the solution of algebraic equations by whole numbers, Diophantine problems, go back to the book of Diophantus in the Hellenistic period in ancient Alexandria. A recent major breakthrough in Diophantine equations, the proof of the Mordell conjecture by the young German mathematician Faltings, asserts that the number of such solutions for each equation of a rather general type must be finite. It is achieved with the application of a broad variety of recently created tools in modern abstract algebra, combined with delicate geometric arguments.

The richer the repertoire of modern mathematical research, the broader the arsenal of concepts and tools available for the use of the mathematicized sciences. The difficulty lies in the problem of communication, of the scientific practitioners being able to penetrate through the difficulties of translation between the languages of different disciplines, of knowing what is relevant among what is available in terms of concept and technique.

As the concerns and principal foci of scientific interest move into domains ever further from the classical domains of theory and experience, the role of mathematical ideas and techniques inevitably grows since they often provide the only tools by which one can probe further into the unknown. This is particularly true for domains involving complexity of organization or nonlinearity of interaction, which I have suggested above constitute the future front line of the major themes of scientific advance. Though they may become the subject-matter of major themes of scientific disciplines in their own right, I doubt that this will lead to the disappearance of professional differences between specialists in various disciplines in attacking these scientific problems. The difference between specialties has a positive function, as well as its negative consequences. Each specialist can rely upon the intellectual traditions and resources of his scientific specialty, and this applies with the greatest force to the mathematician. What we can ask for is a broader and more effective effort at communication among those concerned with common problems and an active interest in and sympathy with the thematic concerns of other specialties than one's own.

It is this effort at communication and sympathetic interrelation that justifies efforts to construct new institutional forms that try to bridge classical disciplinary barriers. The present-day university with its usual kinds of departmental barriers often tends to frustrate communication and dissolve any sense of sympathetic interrelation. Tribal modes of thought in particular disciplines may lead those in one field to regard sympathy with or interest in other fields, particularly in students, as signs of incipient disciplinary treason. In my oral presentation, I cited three paraticular cases of which I knew to illustrate this point in a sharp form, two involving other participants in the Workshop.

Institutes like the proposed Santa Fe Institute cannot by the nature and structure of the present scientific world replace the fundamental role of the research universities. They have another function, to serve as paradigms of alternative modes of organization and action. If successful, they will goad the conventional structure of the research universities into imitating them, and as the proverb goes, imitation is the sincerest form of flattery. The greatest danger to the continued thrust of scientific and mathematical discovery is the possibility that the institutions which house that thrust will become routinized and bureaucratized. Let us hope that the success of the efforts to create innovative new institutions will provide a meaningful countervailing thrust against the deadly threat of the strangulation of science by routine and bureaucracy.

HARVEY FRIEDMAN
Department of Mathematics, Ohio State University, Columbus, OH 43210

Applications of Mathematics to Theoretical Computer Science

I am deeply flattered and at the same time overwhelmed to be invited to give a presentation on a topic of this scope to a group of such eminent scholars. I accepted this invitation not because I possess the qualifications to do justice to the task—I don't—but out of great interest and respect for what you are trying to do. I have worked primarily in mathematical logic and the foundations of mathematics, and only secondarily in theoretical computer science. I will try to do the best I can under the circumstances.

What I have done is to list ten substantial areas of theoretical computer science in which either (a) sophisticated mathematical ideas and constructions have been used to obtain significant results, or (b) the conceptual framework is so closely related to that of an existing developed area in mathematics, that the computer science area can be viewed as almost a redirection of the mathematics area.

Let me describe these ten areas—I'm sure that many more could be discussed—as well as the allied areas of mathematics:

1. ABSTRACT COMPLEXITY AND RECURSION THEORY.

Computational complexity theory is a theoretical study of what idealized computers can or cannot do, under various relevant restrictions of computing resources such as the amount of available time or space. Among the idealized models used in this area have been the "multitape Turing machines" and the "random access machines." More recently, variants of these models involving parallel computation, idealized circuits, and probabilistic algorithms have been studied.

In abstract complexity, basic features of these models and their mutual relationships are studied, independently of how they apply to computing problems of special interest. The conceptual framework is closely akin to that of recursion theory, which is a branch of mathematical logic concerned with what idealized computers can or cannot do, independently of restrictions on computing resources. In fact, recursion theory originated in the 1930's with the advent of these same Turing machines by Alan Turing.

The basic problems in abstract complexity are at this time extremely difficult mathematical problems with clear and attractive formulations. For instance, if an attribute can be tested by an algorithm whose *space* requirements are no more than a polynominal function of the size of the object being tested, then can the attribute be tested by an algorithm whose *time* requirements are no more than a polynomial function of the size of the input? In order to answer such questions definitively, some powerful new mathematical techniques need to be developed.

2. CONCRETE COMPLEXITY AND MATHEMATICS.

Concrete complexity theory is the study of what idealized computers can or cannot do, subject to relevant limitations of resources, in the context of specific problems of special interest. Generally speaking, the results are cast in terms of the same models used in abstract complexity, but when positive results are obtained, the algorithms are often reassessed and modified to run efficiently on actual computers.

As an example, linear programming problem asks whether there is a vector subject to a given set of linear inequalities, and to find one if there is one. The classical Dantzig "simplex method" is quite efficient in practice, but very inefficient in theory. The Kachian "ellipsoid method" is quite efficient in theory, but turned out to be very inefficient in practice. Much sophisticated mathematics has gone into explaining this disparity between theory and practice, with some degree of success. Another chapter in this situation is unfolding with work of Karmarkar which gives another geometric algorithm which is efficient both in theory and practice.

A second example is that of primality testing—testing whether a number is prime. Efficient algorithms are known both from the standpoint of theory and practice. These algorithms rely heavily on classical number theory and some of them involve the Riemann zeta hypothesis.

Concrete complexity is an entirely open-ended area touching virtually every area of mathematics. Essentially every area of mathematics is being restudied from the computational complexity point of view today.

3. AUTOMATA, FORMAL LANGUAGE THEORY, COMPILERS, SEMIGROUPS, AND FORMAL POWER SERIES.

The finite state automata are a very weak kind of abstract machine model which is a basic building block in the detailed construction of both abstract and actual computers. They are, however, the strongest kind of abstract machine model for which we have a reasonably complete structure theory. The semigroups from algebra are used to represent these automata. Methods in mathematics for decomposing semigroups lead to the decomposition of these automata into irreducible component machines from which all machines are built.

Formal language theory is an outgrowth of Chomsky's work in linguistics. Chomsky's categories are suitable in contexts far removed from natural language. Schutzenberger and others have developed a theory relating the context free languages with a branch of algebra called formal power series.

Formal language theory has been an essential ingredient in the specification and construction of compilers.

4. PARALLEL ARCHITECTURES AND PERMUTATION GROUPS.

Almost all of today's computers are based on the "Von Neumann architecture," in which computation proceeds serially. Roughly speaking, at any given time something is happening only at, at most, one location in the computer. General purpose computers based on parallel architectures are now feasible because of advances in hardware. Several of the architectures proposed are based on intriguing mathematical schemes. One of them that has been developed by Jack Schwartz and others is the "shuffle exchange network." A large number of components are linked in pairs like a deck of cards after a perfect shuffle. This creates, in effect, a permutation on the components which can be applied in the course of computation. The fruitfulness of the scheme relies on properties of the group of permutations on a finite set.

Modifications of this idea form the basis of the current ultra-computer project jointly run by IBM and NYU.

5. PROGRAMMING LANGUAGES AND FORMAL SYSTEMS.

It is widely recognized that the principal programming languages—particularly the general purpose languages—leave much to be desired in that it is unexpectedly difficult and time consuming to write programs in them, debug them, and read and understand anyone else's programs. John Backus, the developer of FORTRAN, has been insisting that radically new languages are required to solve this problem, and that the reliance on assignment and control statements are at the heart of the trouble. He has been advocating "functional programming languages" to meet this challenge. These languages are very close in syntax and semantics to the kind of formal systems encountered in mathematical logic. They have no assignment or control statements. The principal difficulty is, at present, the unacceptable loss of efficiency in the implementation of such languages as compared to that of the usual languages. Experience from logic in the construction of concise formal systems with clear semantics (admittedly in contexts other than computation) is expected to play a crucial role in the development of efficient functional programming languages.

6. AUTOMATIC THEOREM PROVING AND PROOF THEORY.

Automatic theorem proving in the sense of automatically developing proofs of interesting mathematical conjectures is a ridiculously over-ambitious goal that is commonly rejected today. However, the more modest goal of using automated deduction in conjunction with mathematicians is being pursued. In a very broad sense, this is already happening: the computer simulations of differential equations, exhaustive tests of cases in the proof of the four-color theorem, etc. In the strict sense, the most promising focus is on automated proof checking. Even in extremely detailed proofs, mathematicians will never fill in all the little routine details. It is up to the proof checker to fill in these gaps. The goal is to write increasingly powerful proof checkers that fill in increasingly larger gaps. The area is closely allied with program verification discussed below.

The whole conceptual framework and methodology of this area is virtually identical with parts of proof theory—a branch of mathematical logic. It really should be regarded as a branch of mathematical logic.

7. PROGRAM VERIFICATION AND PROOF THEORY.

The goal of program verification is to automatically check that a given program is correct, i.e., behaves in the manner intended. This area leans on automatic proof checking. The relevant approach to this depends on the features of the language used to write the to-be-verified programs. There is the problem of the formal specification of the intention of a program, which is sometimes hard to resolve. Ideally, for suitable functional programming languages, the program verification problem should be relatively easy. The framework and methodology of this area is also closely related to parts of proof theory.   .

8. ROBOTICS AND ALGEBRA, GEOMETRY.

Profound applications of algebra and geometry are expected to basic problems of motion planning. For instance, Schwartz and Sharir have recently considered aspects of the "Piano Movers" problem: "that of finding a continuous motion which will take a given body or bodies from a given initial position to a desired final position, but which is subject to certain geometric constraints during the motion." They make sophisticated use of variants of the work of the late logician Alfred Tarski on the quantified elementary theory of real numbers, and there are close connections with real algebraic geometry.

9. PROTOCOLS, SECURITY AND NUMBER THEORY.

One typical problem in this area is the following: suppose we all have identification numbers known only to us and a giant computer in Washington. We wish to mail messages to this Washington computer by mail, and we want to sign the message in the sense that the computer will know the identification number of the person who sent the message. But we do not want our identification number revealed even if our message is intercepted; and we also do not want this message to be understood in the event of interception. Many of the proposed protocols for accomplishing this rely on clever observations from classical number theory, and involve the primality testing mentioned earlier. However, these schemes have never been satisfactorily proved to be secure. To do this, very difficult mathematical problems have to be solved, such as the computational intractability of factorization of integers.

10. DATABASES AND MODEL THEORY.

The main problem in databases is how to maintain a potentially massive amount of data in such a way that information based on this stored data can be retrieved efficiently, and the data can be efficiently updated. A basic goal is to allow the user to make a relatively wide variety of inquiries. The main framework for dealing with these issues is virtually identical to that used in model theory, a branch of mathematical logic. Many of the theoretical problems in this area amount to asking for an (efficient) algorithm for deciding whether a sentence of a special form in ordinary predicate logic is valid. This has always been a standard kind of problem in model theory.

This ends my discussion of these selected ten areas.

Some of you yesterday asked what NP-completeness is. NP stands for "nondeterministic polynomial time." P stands for "deterministic polynomial time."

A set E of strings from a finite alphabet is said to be in P if there is a deterministic Turing machine which accepts exactly the strings in E, and does so in an amount of time that is bounded by a polynomial in the length of the string accepted. Deterministic Turing machines have an obvious genealization to nondeterministic Turing machines, where in any state, reading any symbol, the program gives the machine a finite choice of actions to take. E is then said to be in NP if there is a *nondeterministic* Turing machine which accepts exactly the strings in E, and does so in an amount of time that is bounded by a polynomial in the length of the string accepted. E is said to be NP-complete if every set in NP is polynomial time reducible to it. A massive number of interesting problems turn out to be NP-complete. The open problem is whether P = NP. If any one NP-complete problem is in P, then they all are, i.e., then P = NP.

All of the concepts above are robust in that they are independent of the choice of the machine model (within reasonable limits).

M. P. SCHÜTZENBERGER
Université Paris VII

Linguistics and Computing

The disciplines of linguistics and computer science hold in common the distinction of having to reason, at almost every step, on both SYMBOLS and the MEANING of these symbols. That is to say, they have to stay as close as possible to the rigorously formalistic sciences and at the same time coast along the unfathomable abyss where thought is in danger of getting lost in examination of the meaning of reality and in discussion of intent.

Thus, from the first, computer science has had to face language problems, either within itself in the establishment of codes that allow communication with the machine, or in its translations into computer language or the classification of data. One of the most popular activities in the world of computing today has become the elaboration of the word-processing programs that are gradually replacing our secretaries.

I intend here first to sketch briefly a few historical landmarks in the development of these ideas, and then to give an account of some more recent research that I believe fits in with the aims of our gathering.

In the vast domain of linguistic resarch, phonetics and phonology are the only branches in which the study of SYMBOLS is sufficient in itself. In all others, it is necessary to provide initial guidelines so that the examination of meaning does not intrude on each and every step of reasoning. This is clearly the case in the important branch of classical philology and its recent developments in the grouping and filiation of the languages of the world.

The basic method has been and remains the comparison of vocabularies. On the one hand, it is a purely formal study of the symbols which enable us to perceive that "pater" and "father" are very close in their sequence of phonemes. On the other hand, the fact that these words express the same relationship of consanguinity, that they have the same meaning, requires nothing more than setting up a lexicon. To accomplish this task, it is sufficient to know the most rudimentary semantic equivalence, the same that any speaker acquainted with both languages can deliver instantaneously. Or if you prefer, this equivalence requires only a very small amount of information about the semantic content of the words under study. Thus, to establish a possible relationship between Turkish and Quechua (I take this tragicomic example intentionally), it is sufficient to have a dictionary of Turkish-X and Quechua-X, and it does not matter whether X is English, German, Spanish, or French. This is less trivial than it appears, for experience has shown that there is very little guarantee of the validity of translation from a Turkish text to a Quechua one if it is based on secondary translations of the texts in another language X. As soon as a more subtle meaning is required for the comparison, the central point is in grave danger of getting lost. Such examples abound, even among the highly ritualized languages that are in use at the UN.

On the other hand, in the matter of theorizing about or programming the syntax, linguists have slowly elaborated a system of concepts which enable us more or less to compare the grammar of a language Y with the grammar of a language Z in terms of the categories of a third language X. However, it is probable that the preliminary abstractions are debatable, for the arguments used to define such categories as a verb or an adjective do not in any way come from the previous universal formal logic, which would have to contain the ineffable reality of the vision man has of what action or quality is.

Syntax, style, modes of discourse, and so on, present so many more problems for the resolution of their meaning that I won't even touch on them!

The language of computing started, as you all know, with the most elementary babble: destination orders. Fortran, the first symbolic language, was still quite elementary. And regardless of what my most enthusiastic computer colleagues say, all those who use computing machines know at their own expenses that not too much progress has been made. Except, of course, on the subject of grammar. It is curiously through the work of formal but traditional linguists (I mean Chomsky and his school) that the model of an effective language has been developed to describe the language of programming. I repeat that this is valid for GRAMMAR only (computer experts call it "syntax"). As for semantics, more later.

There have been numerous applications of computing theory to linguistics. Everyone knows the unfortunate fate of automatic translation, in which so much research and talent was sunk without producing any real results, as the Pierce commission showed in its remarkable autopsy report. The same wave brought on the birth of what is known as quantitative linguistics. Besides the problems of copyright—where the advent of computers has enabled one to use in full statistical methods inconceivable without that tool—quantitative linguistics does not seem to have produced new or unexpected results. Could it be, as one of my friends says, that a quantitative computing vertigo has obnubilated the true interest in and knowledge of linguistics of those who venture to play with computers?

The work of M. Gross seems to be worth mentioning here, and I shall briefly summarize it as a transition to the description of some new avenues of research which I think are fruitful.

Gross set out to test on a real scale Chomsky's theory of grammar, which originally prescribed a language model capable of describing the grammatical accuracy of sentences with a formal system that included only a limited number of initial data. This very model, as I said above, has become the basis of the programming syntax. Its validity for natural languages, if one established it, would show that it is possible to isolate a level at which the study of signs is sufficient by itself, for the totality of references to meaning is included in the specific rules for each language.

Attempts at validation tried before Gross—for English, Hebrew, Turkish, and certain Amerindian languages—seemed rather convincing. However, they suffered from a defect, very frequently seen in the so-called applications of computer science: each and every specialist in the above-named languages had proceeded to examine only a very small fraction of their vocabulary.

Gross, a man trained at the tough school of "hard sciences," undertook with the help of a computer to examine one language exhaustively, French. The result of his considerable labor was extremely surprising: every WORD in the French dictionary, or almost every word, requires a special rule, that is to say a specific and complex initial datum, even when all that is sought is only to guarantee the approximate grammatical correctness of sentences, without concern whether they have meaning or not. Furthermore, he was unable to find a reorganization that lightened the task of assembling the data base. In short, it is necessary to conclude that to establish a MINIMAL but somewhat COMPLETE French grammar requires such a huge initial mass of data, that its further use is rather futile, except, of course, for some subsidiary aspects (conjugations and such), the very same which are given by traditional grammars.

This, of course, is not specific to French. The techniques developed for that language have revealed the same phenomenon when applied to English, Spanish, Arabic, and others. A second phenomenon became apparent in the course of the investigation: the abundance of FIXED PHRASES and their role in the transmission of meaning. Gross showed that, using a method of analysis of transformations due to Zellig Harris, those fixed phrases formed a new and important class of linguistic objects. They were what a physicist would call word complexes with high binding energy. These complexes play a qualitative role different from that of words or propositions.

The comparative studies being pursued actively today appear more promising, for they could provide new means to ascertain the similarities and differences between natural languages. Such studies require the computational expertise and international collaboration of specialists with different native languages, for at this deep level of analysis, work not in the researcher's own tongue has been shown to be very certain.

There exist two other avenues of recent research which seem to me to be particularly promising. Both deal with what one could call the internal linguistics of computer science.

The first is semantic. Its aim is to establish concretely useful relations between programs as sequences of abstract symbols and these same objects as sequences of instructions. My friend J. Arsac, to whom I am much indebted for all that relates to the dialectic of symbols and meaning, has discovered certain rules for the formal transformations of the sequences while retaining their machine interpretation. The conditions of their use are, of course, relatively limited, but the algorithms that bring them about have been applied experimentally to programs that already exist. The application of these automatic transformations has allowed substantial gains in time (of the order of ten to thirty percent in the majority of cases) for program of average size written by average programmers.

It would be very important to devise the means of extending these methods to supercomputers employing vectorization, for the optimizers actually in use are far from perfect. By the same token, this automatic application of man-made programs to the demands of the machines' efficiency would enable one to improve both languages and systems from an ergonomic point of view, that is, from the point of view of the user.

Such endeavors are being undertaken in many centers, and their evocation here most certainly has not revealed anything new to those of you who follow the advances of computing science. I have outlined them because they constitute an island of precise work in an ocean of theoretical research whose applicability is ever pushed back into an uncertain future. They use explicitly some mathematical techniques developed since the early days of computing, that is, the theory of words and that of formal languages, or groups of words. M. Lothaire and S. Eilenberg have written about the state of the art as it was a few years ago. Since then, more recent studies have shown the importance of certain chapters (such as those on the theory of "infinite words") in the study of problems of synchronization and parallelism in the most concrete aspects of computing.

It is on this last example of the bond between a formal theory of symbols and its significant applications that I conclude my exposé.

CHARLES H. BENNETT
IBM Research, Yorktown Heights, NY 10598, April 1985

Dissipation, Information, Computational Complexity and the Definition of Organization

I address two questions belonging to an interdisciplinary area between statistical mechanics and the theory of computation:

1. What is the proper measure of intrinsic complexity to apply to states of a physical system?
2. What role does thermodynamic irreversibility play in enabling systems to evolve spontaneously toward states of high complexity?

I. INTRODUCTION

A fundamental problem for statistical mechanics is to explain why dissipative systems (those in which entropy is continually being produced and removed to the surroundings) tend to undergo "self-organization," a spontaneous increase of structural complexity, of which the most extreme example is the origin and evolution of life. The converse principle, namely that nothing very interesting is likely to happen in a system at thermal equilibrium, is reflected in the term "heat death." In the modern world view, thermodynamic driving forces, such as the temperature difference between the hot sun and the cold night sky, have taken over one of the functions of God: they make matter transcend its clod-like nature and behave instead in dramatic and unforseen ways, for example molding itself into thunderstorms, people, and umbrellas.

The notion that dissipation begets self-organization has remained informal, and not susceptible to rigorous proof or refutation, largely through lack of an adequate mathematical definition of organization. Section II, after reviewing alternative definitions, proposes that organization be defined as "logical depth," a notion based on algorithmic information and computational time complexity. Informally, logical depth is the number of steps in the deductive or causal path connecting a thing with its plausible origin. The theory of computation is invoked to formalize this notion as the time required by a universal computer to compute the object in question from a program that could not itself have been computed from a more concise program.

Having settled on a definition of organization, we address briefly in section III the problem of characterizing the conditions (in particular, thermodynamic irreversibility) under which physical systems evolve toward states of high organization. We do not solve this problem, but rather suggest that it can be reduced to several other problems, some of which can already be regarded as solved, some of which are promising areas of research, and some of which are well-known unsolved problems in mathematics (notably the P=PSPACE question).

II. THE PROBLEM OF DEFINING ORGANIZATION

Just what is it that distinguishes an "organized" or "complex" structure like the human body from, say, a crystal or a gas? Candidates for a definition of organization can be divided into those based on function and those based on structure.

A. FUNCTIONAL DEFINITIONS

Living organisms are noted for their capacity for complex function in an appropriate environment, in particular the ability to grow, metabolize, reproduce, adapt, and mutate. While this functional characterization may be a good way to define "life," in distinction to nonliving phenomena that possess some but not all of life's attributes (e.g., a crystal's trivial growth; a flame's metabolism), it is not really a satisfactory way to define organization. We should still like to be able to call organized such functionally inert objects as a frozen human body, a printout of the human genome, or a car with a dead battery. In other words, what we need is not a definition of life or organism (probably inherently fuzzy concepts anyway), but rather a definition for the kind of structural complexity that in our world is chiefly found in living organisms and their artifacts, a kind that can be produced to a lesser degree by laboratory experiments in "self-organization," but which is absent from such structurally trivial objects as gases and crystals.

Another functional characterization of complexity, more mathematical in flavor than the lifelike properties mentioned above, is as the capacity for universal computation. A computationally universal system is one that can be programmed, through its initial conditions, to simulate any digital computation. For example, the computational universality of the well-known deterministic cellular automaton of Conway called the "game of life" implies that one can find an initial configuration that will evolve so as to turn a certain site on if and only if white has a winning strategy at chess, another initial configuration that will do so if and only if the millionth decimal digit of pi is a 7, and so on. On a grander scale, one can in principle find initial conditions enabling the Conway automaton to simulate any physical or chemical process that can be digitally simulated, even presumably the geological and biological evolution of the earth.

The property of computational universality was originally demonstrated for irreversible, noiseless systems such as Turing machines and deterministic cellular automata having little resemblance to the systems ordinarily studied in mechanics and statistical mechanics. Later, some reversible, deterministic systems (e.g., the hard sphere gas [Fredkin-Toffoli, 1982] with appropriate initial and boundary conditions, and Margolus' billiard ball cellular automaton [Margolus, 1984] which models this gas) have been shown to be computationally universal. Very recently [Gacs, 1983; Gacs-Reif, 1985], certain irreversible, noisy systems (probabilistic cellular automata in 1 and 3 dimensions with all local transition probabilities positive) have been shown to be universal. Computational universality, therefore, now appears to be a property that realistic physical systems can have; moreover, if a physical system does have that property, it is by definition capable of behavior as complex as any that can be digitally simulated.

However, computational universality is an unsuitable complexity measure for our purposes because it is a functional property of systems rather than a structural property of states. In other words, it does not distinguish between a system merely capable of complex behavior and one in which the complex behavior has actually occurred. The complexity measure we will ultimately advocate, called logical depth, is closely related to the notion of universal computation, but it allows complexity to increase as it intuitively should in the course of a "self-organizing" system's time development.

B. THERMODYNAMIC POTENTIALS

In spite of the well-known ability of dissipative systems to lower their entropy at the expense of their surroundings, flouting the spirit of the second law while they obey its letter, organization cannot be directly identified with thermodynamic potentials such as entropy or free energy: the human body is intermediate in entropy between a crystal and a gas; and a bottle of sterile nutrient solution has higher free energy, but lower subjective organization, than the bacterial culture it would turn into if inocculated with a single bacterium.

This difference in free energy means that, even without the seed bacterium, the transformation from nutrients to bacteria (albeit an improbable case of spontaneous biogenesis) is still vastly more improbable case of spontaneous biogenesis) is still vastly more probable than the reverse transformation, from bacteria to sterile, high free-energy nutrients. The situation is analogous to the crystallization of a long-lived supersaturated solution: although crystallization without the catalytic assistance of a seed crystal may be so slow as to be unobservable in practice, it is not thermodynamically forbidden, and is, in fact, overwhelmingly more probable than the reverse process.

Subjective organization seems to obey a "slow growth law" which states that, except by a lucky accident, organization cannot increase quickly in any deterministic or probabilistic process, but it can increase slowly. It is this law which forbids sterile nutrient from turning into bacteria in the laboratory, but allows a similar transformation over geological time. If the slow growth law is to be obeyed, the rapid multiplication of bacteria after inocculation must not represent much increase in organization, beyond that already present in the seed bacterium. This, in turn, means that subjective organization is not additive: 1 bacterium contains much more organization that 0 bacteria, but 2 sibling bacteria contain about the same amount as 1.

C. INFORMATION CONTENT

The apparent non-additivity of "organization" suggest another definition for it, namely as information content, an object's information content being the number of bits required to specify it uniquely. Clearly, two large message-like objects (e.g., DNA molecules), if they happen to be identical, do not together contain significantly more information than one alone.

This subsection will review various definitions of information, especially the algorithmic definition implied by the phrase "number of bits necessary to specify a structure uniquely." However, it should be pointed out that information in this sense, like entropy, leads to absurd conclusions when used as the measure of subjective organization: just as the human body is intermediate in entropy between a crystal and a gas, so the human genome is intermediate in information between a totally redundant sequence, e.g., AAAAA..., of near zero information content and a purely random sequence of maximal information content. Although information itself is a poor measure of organization, it will be discussed at some length because it underlies two of the more adequate organization measures to be discussed later, vis. mutual information and logical depth.

There is some uncertainty as to how the "information content" of biological molecules ought to be defined. The easiest definition is simply as the information *capacity* of the molecule, e.g., 2N bits for a DNA molecule of N nucleotides. This definition is not very useful, since it assigns all sequences of a given length the same information content.

In the classical formulation of Shannon, information is an essentially statistical property. The information content in bits of a message is defined as the negative base-2 logarithm of its probability of having been emitted by some source, and it is improper to treat information content as if it were a function of the message itself, without specifying the probability. This is rather awkward in a biological context, where one is frequently faced with a bare message, e.g., a DNA sequence, without any indication of its probability. The information capacity is equivalent to assuming a uniform probability distribution over all sequences. It would be more informative to define the information content of a sequence x as its -log probability in some physically specified distribution, such as an (equilibrium or nonequilibrium) statistical mechanical ensemble. However, this approach departs from the goal of making the definition of organization intrinsic to the sequence.

A third approach to defining information is as the number of bits necessary to uniquely describe an object in some absolute sense, rather than with respect to a particular probability distribution. This approach has been put on a firm mathematical basis by regarding the digital object x as the output of a universal computer (e.g., a universal Turing machine), and defining its algorithmic information content $H(x)$ as the number of bits in its "minimal algorithmic description" $x*$, where $x*$ is the smallest binary input string that causes the universal computer to produce exactly x as its output. Clearly this definition depends on the choice of universal computer, but this arbitrariness leads only to an additive $O(1)$ uncertainty (typically $\pm$ a few thousand bits) in the value of $H(x)$, because of the ability of universal machines to simulate one another. Algorithmic information theory also allows randomness to be defined for individual strings: a string is called "algorithmically random" if it is incompressible, i.e., if its minimal description is about the same size as the string itself. Algorithmic information is discussed further in the introductory article by Chaitin [1975], and in review articles by Zvonkin and Levin [1970] and Chaitin [1977].

The advantage of using a universal computer to regenerate the message is that, for sufficiently long messages, it subsumes all other more specialized schemes of effective description and data compression, e.g., the use of a dictionary of abbreviated encodings for frequently occurring subsequences. Any non-universal scheme of data compression fails to compress some sequences of obviously low information content. For example, the sequence consisting of the first million digits of pi, though it admits a concise algorithmic description, probably cannot be significantly compressed by abbreviating frequent sequences.

As noted above, information per se does not provide a good measure of organization, inasmuch as messages of maximal information content, such as those produced by coin tossing, are among the least organized subjectively. Typical organized objects, on the other hand, precisely because they are partially constrained and determined by the need to encode coherent function or meaning, contains less information than random sequences of the same length; and this information reflects not their organization, but their residual randomness.

For example, the information content of a genome, as defined above, represents the extent to which it is underdetermined by the constraint of viability. The existence of noncoding DNA, and the several percent differences between proteins performing apparently identical functions in different species, make it clear that a sizable fraction of the genetic coding capacity is given over to transmitting such "frozen accidents," evolutionary choices that might just as well have been made otherwise.

D. MUTUAL INFORMATION AND LONG-RANGE ORDER

A better way of applying information theory to the definition of organization is suggested by the nonadditivity of subjective organization. Subjectively organized objects generally have the property that their parts are correlated: two parts taken together typically require fewer bits to describe than the same two parts taken separately. This difference, the *mutual information* between the parts, is the algorithmic counterpart of the non-additivity of statistical or thermodynamic entropy between the two parts. In many contexts, e.g., communication through a noisy channel, the mutual information between a message and something else can be viewed as the "meaningful" part of the message's information, the rest being meaningless information or "noise."

A body is said to have long-range order if even arbitrarily remote parts of it are correlated. However, crystals have long-range order but are not subjectively very complex. Organization has more to do with the *amount* of long-range correlation, i.e., the number of bits of mutual information between remote parts of the body. Although we will ultimately recommend a different organization measure (logical depth), remote mutual information merits some discussion, because it is characteristically formed by nonequilibrium processes, and can apparently be present only in small amounts at thermal equilibrium. Notions similar to mutual information have been introduced in many discussions of biological organization, but often without clearly distinguishing among gross information content (i.e., accidental or arbitrary aspects of the object as a whole), mutual information (amount of correlation between parts that individually are accidental and arbitrary), and determined, non-accidental aspects of the object as a whole which, as argued above, are not information at all, but rather a form of redundancy.

If two cells are taken from opposite ends of a multicellular organism, they will have a large amount of mutual information, if for no other reason than the presence in each cell of the same genome with the same load of frozen accidents. As indicated earlier, it is reasonably certain that at least several percent of the coding capacity of natural genomes is used to transmit frozen accidents, and, hence, that the mutual information between parts of a higher organism is at least in the hundred megabit range. More generally, mutual information exists between remote parts of an organism (or a genome, or a book) because the parts contain evidence of a common, somewhat accidental history, and because they must function together in a way that imposes correlations between the parts without strictly determining the structure of any one part. An attractive feature of remote mutual information for physical systems is that it tends to a finite limit as the fineness of coarse graining is increased, unlike simple information or entropy in a classical system.

Since mutual information arises when an accident occurring in one place is replicated or propagated to another remote place, its creation is an almost unavoidable side effect of reproduction in a probabilistic environment. Another obvious connection between mutual information and biology is the growth of mutual information between an organism and its environment when the organism adapts or learns.

Further support for remote mutual information as an organization measure comes from the fact that systems stable at thermal equilibrium, even those with long-range order, exhibit much less of it than nonequilibrium systems. Correlations in systems at equilibrium are generally of two kinds: short-range correlations involving a large number of bits of information (e.g., the frozen-in correlations between adjacent lattice planes of an ice crystal, or the instantaneous correlations between atomic positions in adjacent regions of any solid or liquid), and long-range correlations involving only a few bits of information. These latter include correlations associated with conserved quantities in a canonical or microcanonical ensemble (e.g., if one half of a gas cylinder contains more than half the atoms, the other half will contain fewer than half of the atoms) and correlations associated with order parameters such as magnetization and crystal lattice orientation. In either case, the amount of mutual information due to long-range correlations is small: for example, in a gas of 10^{23} atoms, conservation of the number of atoms causes the entropy of the whole to be about $\log\sqrt{10^{23}} \approx 39$ bits less than the sum of the entropies of its halves. It may at first seem that a real-valued order parameter, such as phase or orientation of a crystal lattice, already represents an infinite amount of information; however, in an N-atom crystal, owing to thermal and zero-point fluctuations, the instantaneous microstate of the entire crystal suffices to determine such order parameters only to about $\log N$ bits precision; and, hence, the mutual information between remote regions of a macroscopic crystal amounts to only a few dozen bits.

Unfortunately, some subjectively not-very-organized objects also contain large amounts of remote mutual information. For example, consider an igneous rock or other polycrystalline solid formed under nonequilibrium conditions. Such solids, though not subjectively very "organized," typically contain extended crystal defects such as dislocations and grain boundaries, which presumably carry many bits of information forward from the earlier-crystallized to the later-crystallized portions of the specimen, thus giving rise to the correlated frozen accidents that constitute mutual information. On a larger scale, terrestrial and planetary geological processes create large amounts of mutual information in the form of complementary fracture surfaces on widely separated rock fragments. Mutual information does not obey the slow growth law, since an ordinary piece of glass, after a few minutes of hammering and stirring, would be transformed into a three-dimensional jigsaw puzzle with more of it than any genome or book. Even larger amounts of mutual information could be produced by synthesizing a few grams of random, biologically meaningless DNA molecules, replicating them enzymatically, and stirring the resulting mixture to produce a sort of jigsaw-puzzle soup. Two spoonfuls of this soup would have macroscopically less than twice the entropy of one spoonful. In all these examples, the mutual information is formed by nonequilibrium processes and would decay if the system were allowed to approach a state of true thermal equilibrium, e.g., by annealing of the separated fracture surfaces. Remote mutual information is somewhat unsatisfying as a measure of organization because it depends on accidents, assigning low organization to some objects (such as the binary expansion of pi) which seem organized though they lack accidents, and high organization to other objects whose correlated accidents are of a rather trivial sort (random palindromes, broken glass).

E. SELF-SIMILARITY

A conspicuous feature of many nontrivial objects in nature and mathematics is the possession of a fractal or self-similar structure, in which a part of the object is identical to, or is described by the same statistics as, an appropriately scaled image of the whole. I feel that this often beautiful property is too specialized to be an intuitively satisfactory criterion of organization because it is absent from some intuitively organized objects, such as the decimal expansion of pi, and because, on the other hand, self-similar structures can be produced quickly, e.g., by deterministic cellular automata, in violation of the slow growth law. Even so, the frequent association of self-similarity with other forms of organization deserves comment. In some cases, self-similarity is a side-effect of computational universality, because a universal computer's ability to simulate other computers gives it, in particular, the ability to simulate itself. This makes the behavior of the computer on a subset of its input space (e.g., all inputs beginning with some prefix p that tells the computer to simulate itself) replicate its behavior on the whole input space.

F. LOGICAL DEPTH

The problem of defining organization is akin to that of defining the value of a message, as opposed to its information content. A typical sequence of coin tosses has high information content, but little message value; an ephemeris, giving the positions of the moon and planets every day for a hundred years, has no more information than the equations of motion and initial conditions from which it was calculated, but saves its owner the effort of recalculating these positions. The value of a message, thus, appears to reside not in its information (its absolutely unpredicatble parts), nor in its obvious redundancy (verbatim repetitions, unequal digit frequencies), but rather in what might be called its buried redundance—parts predictable only with difficulty, things the receiver could in principle have figured out without being told, but only at considerable cost in money, time or computation. In other words, the value of a message is the amount of mathematical or other work plausibly done by its originator, which its receiver is saved from having to repeat.

Of course, the receiver of a message does not know exactly how it originated; it might even have been produced by coin tossing. However, the receiver of an obviously non-random message, such as the first million bits of pi, would reject this "null" hypothesis on the grounds that it entails nearly a million bits worth of ad-hoc assumptions, and would favor an alternative hypothesis that the message originated from some mechanism for computing pi. The plausible work involved in creating a message, then, is the amount of work required to derive it from a hypothetical cause involving no unnecessary ad-hoc assumptions.

These ideas may be formalized in terms of algorithmic information theory: a message's most plausible cause is identified with its minimal algorithmic description, and its "logical depth," or plausible content of mathematical work, is (roughly speaking) identified with time required to compute the message from this minimal description. Formulating an adequately robust quantitative definition of depth is not quite this simple and, in particular, requires a properly weighted consideration of other descriptions besides the minimal one. When these refinements are introduced [cf Appendix], one obtains a definition of depth that is machine independent, and obeys the slow growth law, to within a polynomial depending on the universal machine. The essential idea remains that a deep object is one that is implausible except as the result of a long computation.

It is a common observation that the more concisely a message is encoded (e.g., to speed its transmission through a channel of limited bandwidth), the more random it looks and the harder it is to decode. This tendency is carried to its extreme in a message's minimal description, which looks almost completely random (if $x*$ had any significant regularity, that regularity could be exploited to encode the message still more concisely) and which, for a nontrivial (deep) message, requires as much work to decode as plausibly went into producing the message in the first place. The minimal description $x*$, thus, has all the information of the original message x, but none of its value.

Returning to the realm of physical phenomena, we advocate identifying subjective organization or complexity with logical depth, in other words, with the length of the logical chain connecting a phenomenon with a plausible hypothesis explaining it. The use of a universal computer frees the notion of depth from excessive dependence on particular physical processes (e.g., prebiotic chemistry) and allows an object to be called deep only if there is no shortcut path, physical or non-physical, to reconstruct it from a concise description. An object's logical depth may, therefore, be less than its chronological age. For example, old rocks typically contain physical evidence (e.g., isotope ratios) of the time elapsed since their solidification, but would not be called deep if the aging process could be recapitulated quickly in a computer simulation. Intuitively, this means that the rocks' plausible history, though long in time, was rather uneventful, and, therefore, does not deserve to be called long in a logical sense.

The relevance of logical depth to physical self-organization depends on the assumption that the time development of physical systems can be efficiently simulated by digital computation. This is a rather delicate question; if by simulation one means an exact integration of differential equations of motion, then no finite number of digital operations could simulate even one second of physical time development. Even when simulation is defined less restrictively (roughly, as an effective uniformly convergent approximation by rational numbers), Myhill [1971] showed that there is a computable differentiable function with a noncomputable solution. On the other hand, it remains plausible that realistic physical systems, which are subject throughout their time development to finite random influences (e.g., thermal and gravitational radiation) from an uncontrolled environment, can be efficiently approximated by digital simulation to within the errors induced by these influences. The evidence supporting this thesis is of the same sort, and as strong as, that supporting the empirically very successful master equation [van Kampen, 1962], which approximates the time development of a statistical mechanical system as a sequence of probabilistic transitions among its coarse-grained microstates.

Accepting the master equation viewpoint, the natural model of physical time development, at least in a system with short-ranged forces, would be a three-dimensional probabilistic cellular automaton. Such automata can be simulated in approximately linear time by a universal three-dimensional cellular automaton each of whose sites is equipped with a coin-toss mechanism; hence, time on such a universal automaton might be the most appropriate dynamic resource in terms of which to define depth. Usually we will be less specific, since other reasonable machine models (e.g., the universal Turing machines in terms of which algorithmic information theory is usually developed) can simulate probabilistic cellular automata, and one another, in polynomial time. We will assume conservatively that any t seconds in the time development of a realistic physical system with N degress of freedom can be simulated by probabilistic computation using time bounded by a polynomial in Nt.

Although time (machine cycles) is the complexity measure closest to the intuitive notion of computation work, memory (also called space or tape) is also important because it corresponds to a statistical mechanical system's number of particles or degrees of freedom. The maximum relevant time for a system with N degrees of freedom is of order $2^{O(N)}$, the Poincaré recurrence time; and the deepest state such a system could relax to would be one requiring time $2^{O(N)}$, but only memory N, to compute from a concise description.

Unfortunately, it is not known that any space-bounded physical system or computer can indeed produce objects of such great depth (exponential in N). This uncertainty stems from the famous open P=?PSPACE question in computational complexity theory, i.e., from the fact that it is not known whether there exist computable functions requiring exponentially more time to compute than space. In other words, though most complexity theorists suspect the contrary, it is possible that the outcome of every exponentially long computation or physical time evolution in a space-bounded system can be predicted or anticipated by a more efficient algorithm using only polynomial time.

A widely held contrary view among complexity theorists today, considerably stronger than the mere belief that P is not equal to PSPACE, is that there are "cryptographically strong" pseudorandom number generators [Blum-Micali, 1984; Levin, 1985], whose successive outputs, on an N-bit seed, satisfy all polynomial time (in N) tests of randomness. The existence of such generators implies that space-bounded universal computers, and, therefore, any physical systems that mimic such computers, can after all produce exponentially deep outputs.

If, on the other hand, it turns out that P=PSPACE, then exponentially deep N-bit strings can still be produced (by well-known "diagonal" method, the gist of which is to generate a complete list of all shallow N-bit strings and then output the first N-bit string not on the list), but the computations leading to these deep objects will require more than polynomial space during their intermediate stages.

It is worth noting that neither algorithmic information nor depth is an effectively computable property. This limitation follows from the most basic result of computability theory, the unsolvability of the halting problem, and reflects the fact that although we can prove a string nonrandom (by exhibiting a small program to compute it), we can not, in general, prove it random. A string that seems shallow and random might, in fact, be the output of some very slow-running, small program, which ultimately halts, but whose halting we have no means of predicting. This open-endedness is also a feature of the scientific method: a phenomenon that seems to occur randomly (e.g., pregnancy) may later turn out to have a cause so remote or unexpected as to have been overlooked at first. In other words, if the cause of a phenomenon is unknown, we can never be sure that we are not underestimating its depth and overestimating its randomness.

The uncomputability of depth is no hindrance in the present theoretical setting where we assume a known cause (e.g., a physical system's initial conditions and equations of motion) and try to prove theorems about the depth of its typical effects. Here, it is usually possible to set an upper bound on the depth of the effect by first showing that the system can be simulated by a universal computer within a time t and then invoking the slow growth rule to argue that such a computation, deterministic or probabilistic, is unlikely to have produced a result much deeper than t. On the other hand, proving lower bounds for depth, e.g., proving that a given deterministic or probabilistic cause certainly or probably leads to a deep effect, though always possible in principle, is more difficult, because it requires showing that no equally simple cause could have produced the same effect more quickly.

III. TOWARDS AN UNDERSTANDING OF THE NECESSARY AND SUFFICIENT CONDITIONS FOR SELF-ORGANIZATION

We have already pointed out a mathematical requirement, namely the conjectured inequality of the complexity classes P and PSPACE, necessary for a finite model system to evolve to a state of depth comparable to its Poincare time. In this section, we mention recent results in computation theory and statistical mechanics which may soon leads to a comprehensive understanding of other conditions necessary and sufficient for model systems to self-organize, i.e., to evolve deterministically or with high probability to a state's deep compared to the system's initial condition.

It is clear that universal computation, and, hence, self-organization, can occur *without dissipation* in reversible deterministic systems such as Fredkin and Toffoli's "billiard ball model" [1982], which consists of classical hard spheres moving on a plane with fixed obstacles (without loss of generality the array of obstacles may be taken to be spatially periodic); or in Margolus' billiard ball cellular automaton [1984] which discretely simulates this model. In these models, the initial condition must be low-entropy, because a reversible system cannot decrease its own entropy (the continuous billiard ball model, because of the dynamical instability of its collisions, in fact requires an initial condition with infinite negative entropy relative to the random hard sphere gas). Moreover, if the system is to preform a nontrivial computation, the initial condition must lack translational symmetry, because a deterministic system cannot break its own symmetries. It would suffice for the initial condition to be periodic except at a single site, which would serve as the origin for a depth-producing computation.

The systems just considered are noiseless. As indicated earlier, it is more realistic to imagine that a physical system is subject to environmental noise, and to treat its motion as random walk, rather than a deterministic trajectory, on the relevant discrete or continuous state space.

In general, such noisy systems require at least some dissipation to enable them to correct their errors and engage in a purposeful computation; the amount of dissipation depends on the noise's intensity and especially on its pervasiveness, i.e., on whether it is considered to affect all, or only some aspects of the system's structure and operation. At the low end of the pervasiveness spectrum are systems such as the clockwork computer of Bennett [1982], in which the noise causes only transitions forward and backward along the intended path of computation, not transitions from one computation into another, or transitions that degrade the structure of the hardware itself. In such systems, all errors are recoverable and the required dissipation tends to zero in the limit of zero speed. More pervasive noise can be found in the situation of error-correcting codes, where some unrecoverable errors occur but the decoding apparatus itself is considered perfectly reliable; and in proofreading enzyme systems [cf Bennett, 1979], where the decoding apparatus is unreliable but still structurally stable. These systems require finite dissipation even in the limit of zero speed. Von Neumann's [1952] classic construction of a reliable computer from unreliable parts is also of this sort: all gates were considered unreliable, but the wires connecting them were considered reliable and their complex interconnection pattern structurally stable. Only recently has decisive progress been made in understanding systems at the high end of the pervasiveness spectrum, in particular, "noisy" cellular automata (henceforth NCA) in which all local transition probabilities are strictly positive. For such an automaton, any two finitely differing configurations are mutually accessible.

An NCA may be synchronous or asynchronous, reversible or irreversible. The former distinction (i.e., between a random walk occurring in discrete time or continuous time) appears to have little qualitative effect on the computing powers of the automata, but the latter distinction is of major importance. In particular, irreversible NCA can function as reliable universal computers [Gacs, 1983; Gacs-Reif, 1985], and can do so robustly despite arbitrary small perturbations of their transition probabilities; while reversible NCA, for almost all choices of the transition probabilities, are ergodic, relaxing to a structurally simple state (the thermodynamic phase of lowest free energy) independent of the initial condition. Irreversibility enables NCA to be robustly nonergodic essentially by protecting them from the nucleation and growth of a unique phase of lowest free energy [Toom, 1980; Domany-Kinzel, 1984; Bennett-Grinstein, 1985].

(An NCA is considered reversible or nondissipative if its matrix of transition probabilities is of the "miscroscopically reversible" form DSD^{-1}, where D is diagonal and S symmetric. In that case, a movie of the system at equilibrium would look the same shown forwards as backwards and the stationary distribution can be represented (exactly for asynchronous automata, approximately for synchronous) as the Boltzmann exponential of a locally additive potential. On the other hand, if the local transition probabilities are not microscopically reversible, the stationary macrostate is dissipative (corresponding physically to a system whose environment continually removes entropy from it), a movie of the system would not look the same forwards as backwards, and the distribution of microstates, in general, cannot be approximated by the exponential of any locally additive potential. Asynchronous reversible NCA, otherwise known as generalized kinetic Ising models, are widely studied in statistical mechanics.)

The computationally universal NCA of Gacs and Gacs-Reif are still somewhat unsatisfactory because they require special initial conditions to behave in a nontrivial manner. A truly convincing case of self-organization would be an NCA with generic transition probabilities that would initiate a depth-producing computation from generic initial conditions (e.g., a random soup). Such an automaton has not been found, though Gacs believes it can be. If it is found, it will lend support to the philosophical doctrine that the observed complexity of our world represents an intrinsic propensity of nature, rather than an improbable accident requiring special initial conditions or special laws of nature, which we observe only because this same complexity is a necessary condition for our own existence.

APPENDIX: MATHEMATICAL CHARACTERIZATION OF DEPTH

Two rather different kinds of computing resources have been considered in the theory of computational complexity: static or definitional resources such as program size, and dynamic resources such as time and memory. Algorithmic information theory allows a static complexity or information content to be defined both for finite and for infinite objects, as the size in bits of the smallest program to computer the object on a standard universal computer. This minimal program has long been regarded as analogous to the most economical scientific theory able to explain a given body of experimental data. Dynamic complexity, on the other hand, is usually considered meaningful only for infinite objects such as functions or sets, since a finite object can always be computed or recognized in very little time by means of a table look-up or print program, which includes a verbatim copy of the object as part of the program.

In view of the philosophical significance of the minimal program, it would be natural to associate with each finite object the cost in dynamic resources of reconstructing it from its minimal program. A "deep" or dynamically complex object would then be one whose most plausible origin, via an effective process, entails a lengthy computation. (It should be emphasized that just as the plausibility of a scientific theory depends on the economy of its assumptions, not on the length of the deductive path connecting them with observed phenomena, so the plausibility of the minimal program, as an effective "explanation" of its output, does not depend on its cost of execution.) A qualitative definition of depth is quoted by Chaitin [1977], and related notions have been independently introduced by Adleman [1979] ("potential") and Levin [Levin and V'jugin, 1977] ("incomplete sequence").

In order for depth to be a useful concept, it ought to be reasonably machine-independent, as well as being stable in the sense that a trivial computation ought not to be able to produce a deep object from a shallow one. In order to achieve these ends, it is necessary to define depth a little more subtly, introducing a significance parameter that takes account of the realtive plausibility of all programs that yield the given object as output, not merely the minimal program. Several slightly different definitions of depth are considered below; the one finally adopted calls an object "d-deep with b bits significance" if all self-delimiting programs to compute it in time d are algorithmically compressible (expressible as the output of programs smaller than themselves) by at least b bits. Intuitively this implies that the "null" hypothesis, that the object originated by an effective process of fewer than d steps, is less plausible than a sequence of coin tosses beginning with b consecutive tails.

The difficulty with defining depth as simply the run time of the minimal program arises in cases where the minimal program is only a few bits smaller than some much faster program, such as a print program, to compute the same output x. In this case, slight changes in x may induce arbitrarily large changes in the run time of the minimal program, by changing which of the two competing programs is minimal. This instability emphasizes the essential role of the quantity of buried redundancy, not as a measure of depth, but as a certifier of depth. In terms of the philosophy-of-science metaphor, an object whose minimal program is only a few bits smaller than its print program is like an observation that points to a nontrivial hypothesis, but with only a low level of statistical confidence.

We develop the theory of depth using a universal machine U, similar to that described in detail by Chaitin [1975B], which has two tapes, a program tape and work tape. The expression $U(s) = x$ will be used to indicate that the machine, started with the binary string s on its program tape and a blank work tape, embarks on a computation that halts after a finite number of steps, leaving the output x on the work tape. The number of steps (run time) is denoted $t(s)$. The work tape can also be used as an auxiliary input, with $U(s, w)$ denoting the output and $t(s, w)$ the run time of a computation beginning with s on the program tape and w on the work tape. In case the computation fails to halt, the functions U and t are considered to be undefined.

The program tape is treated in a special way [Gacs, 1974; Levin, 1974; Chaitin, 1975] in order to allow a natural relative weighting of programs of different lengths. The details of this treatment are described by Chaitin, but the essential feature is that the machine itself must decide how many bits to read off its program tape, without being guided by any special endmarker symbol. Another way of looking at this is to say that the expression $U(s, w) = x$ means that, if the machine were given w on its work tape and any *infinite* binary sequence beginning with s on its program tape, it would halt with the infinite program. This "self-delimiting" formalism allows the *algorithmic probability* of an output x to be defined in a natural way, as the sum of the negative binary exponentials of the lengths of all programs leading to that output:

$$Pu(x) = \sum_{\{s \, : \, U(s) = x\}} 2^{-|s|}$$

Here $|s|$ denotes the length of the binary string s, regarded as a self-delimiting program for the U machine. (Without the self-delimiting requirement, this sum would, in general, diverge.) An analogous conditional algorithmic probability, $Pu(x/w)$, may be defined for computations that begin with a string w on the work tape. This represents the probability that a program generated by coin tossing would transform string w into string x.

Besides being self-delimiting, the U machine must be *efficiently universal* in the sense of being able to simulate any other self-delimiting Turing machine with additive increase in program size and polynomial increase in time and space. That such machines exist is well known. The *minimal program* for a string x, denoted $x*$, is the least string p such that $U(p) = x$. The algorithmic information or entropy of a string $H(x)$ may be defined either as the size of its minimal program, or the negative base-two logarithm of its algorithmic probability, since it can be shown that the difference between these two quantities is bounded by a constant depending on U but independent of x (this is another advantage of the self-delimiting formalism). A string x is said to be compressible by b bits if its minimal program is b bits shorter than x. Regardless of how compressible their outputs may be, all minimal programs are incompressible to within an $O(1)$ constant depending on the standard machine. (If they were not, i.e., if for some s, $x * *$ were significantly shorter than $x*$, then $x*$ would be undercut in its fole as executing $x * *$.) Finite strings, such as minimal programs, which are incompressible or nearly so are called *algorithmically random*. The above formulation in terms of halting, self-delimiting programs appears the most natural way of defining information content for discrete objects such as integers, binary strings, or Ising microstates.

To adequately characterize a finite string's depth, one must consider both the amount of redundancy and the depth of its burial. Several definitions are given below; the best appears to be to say that a string x is (d, b)-*deep*, or d-*deep with b bits significance*, if

i.   every program to compute s in time $\leq d$ is compressible by at least b bits.

It can be shown that any (d, b)-deep string according to this definition is deep in two other, perhaps more intuitive senses:

ii.   computations running in time $\leq d$ supply less than $1/2^{b+O(1)}$ of the string's algorithmic probability.

iii. the smallest program to compute x in time $\leq d$ is at least $b + O(1)$ bits larger than the minimal program $x*$.

Alternative 2), perhaps the most natural (because it fairly weights all computations leading to x) is very close to the chosen definition, since it can be shown (by a proof similar to that of Chaitin's [1975B] theorem 3.2) that any (d, b)-shallow string (one not (d, b)-deep) receives at least $1/2^{b+O(\log b)}$ of its algorithmic probability form programs running in time $\leq d$. Alternative 1) is favored because it satisfies a sharper slow growth law. Alternative 3), perhaps the most obvious, might seriously overestimate the depth of a string with a great many large fast programs, but no single, small, fast program. Whether such strings exist is not known; if they do exist, they should probably not be called deep, since they have a significant probability of being produced by small, fast-running probabilistic algorithms.

It is obviously desirable that depth obey the slow growth law, i.e., that no fast, simple, deterministic or probabilistic algorithm be able to transform a shallow object into a deep one. With the chosen definition of depth, it is easy to show that this is the case: for any strings w and x, if w is less than (d, b) deep, and the algorithmic probability for U to transform w (furnished as an auxiliary input on the work tape) into x within time t is at least 2^{-k}, then s can be no more than $(d + t + O(1), b + k + O(1))$-deep.

Similarly, depth can be shown to be reasonably machine-independent, in the sense that for any two, efficiently universal, self-delimiting machines, there exists a constant c and a polynomial p such that $(p(d), b + c)$ depth on either machine is a sufficient condition for (d, b) depth on the other.

One may well wonder whether, by defining some kind of weighted average run time, a string's depth may reasonably be expressed as a single number. This may, in fact, be done, at the cost of, in effect, imposing a somewhat arbitrary rate of exchange between the two conceptually very different quantities' run time and program size. Proceeding from alternative definition 2) above, one might try to define a string's average depth as the average run time of all computations contributing to its algorithmic probability. Unfortunately, this average diverges because it is dominated by programs that waste arbitrarily much time. To make the average depth of s depend chiefly on the fastest programs of any given size that compute s, it suffices to use the reciprocal mean reciprocal run time in place of a straight average. The *reciprocal mean reciprocal depth* of a string x is thus defined as

$$d_{rmr}(x) = \left[\sum_{\{s\,:\,U(s)=x\}} 2^{-|s|} \right] \Big/ \left[\sum_{\{s\,:\,U(s)=x\}} (2^{-|s|}/t(s)) \right],$$

In this definition, the various computations that produce x act like parallel resistors, the fast computations in effect short-circuiting the slow ones. Although reciprocal mean reciprocal depth doesn't satisfy as sharp a slow growth law as two-parameter depth (multiplicative rather than additive error in the computation time), and doesn't allow strings to have depth more than exponential in their length (due to the short-circuiting of slower programs, no matter how small, by the print program), it does provide a simple quantitative measure of a strong's nontriviality.

An even rougher, qualitative distinction may be drawn between "deep" and "shallow" strings according to whether their reciprocal mean reciprocal depth is exponential or polynomial in the strings' length, or some other parameter under discussion. This rough dichotomy, in which all merely polynomially-deep strings are called shallow, is justified by the typically polynomial cost for one machine model to simulate another, and the consequent arbitrariness in the definition of computation time.

ACKNOWLEDGEMENTS

These ideas developed over many years with the help of countless discussions, especially with Gregory Chaitin, Rolf Landauer, Peter Gacs, Leonid Levin, Tom Toffoli, Norman Margolus, and Stephen Wolfram.

REFERENCES

L. Adleman (1979), "Time, Space, and Randomness," *MIT Report LCS/TM-131.*

C. H. Bennet (1973), "Logical Reversibility of Computation," *IBM J. Res. Develop* **17**, 525.

C. H. Bennet (1979), "Dissipation Error Tradeoff in Proofreading," *BioSystems* **11**, 85–90.

C. H. Bennett (1982), "The Thermodynamics of Computation—a Review," *International J. of Theoretical Physics* **21**, 905–940.

C. H. Bennett and G. Grinstein (1985), "On the Role of Irreversibility in Stabilizing Complex and Nonergodic Behavior in Locally Interacting Systems," submitted to *Phys. Rev. Letters.*

M. Blum and S. Micali (1984), "How to Generate Cryptographically Strong Sequences of Pseudo-Random Bits," *SIAM J. Comput.* **13**, 850–864.

G. Chaitin (1975A), "Randomness and Mathematical Proof," *Sci. Amer.* **232** (May, 1975), 47–52.

G. Chaitin (1975B), "A Theory of Program Size Formally Identical to Information Theory," *J. Assoc. Comput. Mach.* **22**, 329–340.

G. Chaitin (1977), "Algorithmic Information Theory," *IBM J. Res. Develop* **21**, 350–359, 496.

E. Domany and W. Kinzel (1984), *Phys. Rev. Lett.* **53**, 311.

E. Fredkin and T. Toffoli (1982), "Conservative Logic," *International J. of Theoretical Physics* **21**, 219.

P. Gacs (1974), "On the Symmetry of Algorithmic Information," *Soviet Math Dokl.* **15**, 1477.

P. Gacs (1983), *Technical Report No. 132* (Computer Science Department, University of Rochester); to appear in *J. of Computer and System Science.*

P. Gacs and J. Reif (1985), to appear in *1985 ACM Symposium on the Theory of Computing.*

L. A. Levin (1974), "Laws of Information Conservation (nongrowth) and Aspects of the Foundations of Probability Theory," *Problems of Inf. Transm.* **10**, 206–210.

L. A. Levin and V. V. V'jugin (1977), "Invariant Properties of Informational Bulks," *Springer Lecture Notes in Computer Science,* vol. 53, 359–364.

L. A. Levin (1981), "Randomness Conservation Laws; Information and Independence in Mathematics," unpublished manuscript.

L. A. Levin (1985), to appear in *1985 ACM Symposium on the Theory of Computing.*

A. N. Kolmogorov (1965), "Three Approaches to the Quantitative Definition of Randomness," *Problems of Information Transmission* **1**, 1–7.

N. Margolus (1984), "Physics-Like Models of Computation," *Physica* **10D**, 81–95.

J. Myhill (1971), "A Recursive Function Defined on a Compact Interval and Having a Continuous Derivative that is not Recursive," *Michigan Math J.* **18**, 97–98.

M. Pour-El and I. Richards (1981), "The Wave Equation with Computable Initial Data such that its Unique Solution is not Computable," report (U. of Minnesota, School of Mathematics).

R. Solomonoff (1964), "A Formal Theory of Inductive Inference," *Inf. and Contr.* **7**, 1–22 and 224–254.

A. L. Toom (1980), in *Adv. in Probability 6: Multicomponent Systems*, ed. R. L. Dobrushin (New York: Dekker), 549–575.

N. G. van Kampen (1962), "Fundamental Problems in the Statistical Mechanics of Irreversible Processes," *Fund. Prob. in Stat. Mech.*, ed. E. G. D. Cohen (North-Holland).

J. von Neumann (1952), "Probabilistic Logics and the Synthesis of Reliable Organisms from Unreliable Components," *Automata Studies*, ed. Shannon and McCarthy (New Jersey: Princeton University Press).

A. K. Zvonkin and L. A. Levin (1970), "The Complexity of Finite Objects and the Development of the Concepts of Information and Randomness by Means of the Theory of Algorithms," *Russ. Math. Surv.* **256**, 83–124.

GEORGE A. COWAN
Senior Fellow, Los Alamos National Laboratory, Los Alamos, NM

Plans for the Future

These proceedings illustrate and support the observation that many of the most important and challenging activities at the forefront of research range broadly across the conventional disciplines and that, viewed as a whole, such topics represent emerging syntheses in science which may be recognized eventually as new disciplines. Our informal discussions, which have been taped but not summarized in the proceedings, have examined the basis for our concern that these syntheses are frequently poorly defined and nurtured and that new academic options, including the Institute described here, are urgently needed to further define and expedite research in these fields. We have asked some more detailed questions; for example, how do we initially choose staff; how should we rank the emerging syntheses in defining initial programs; what form of governance is desirable during the formative years; and how must it be modified with time and growth?

Our discussions have produced agreement that a number of barriers impede the recognition, support, and pursuit of research at the boundaries between disciplines and that the innovations proposed by the Santa Fe Institute should help lower these barriers. We have agreed that our first priority in organizing the permanent Institute must be on recruiting first-rate people. A major part of the permanent staff and the students must possess the breadth of interest necessary to pursue research on a large number of highly complex and interactive systems which can be properly studied only in an interdisciplinary environment. A ranking of themes will occur naturally as such people are recruited. We have further agreed that education, largely centered on research on these themes, must be our major concern.

The most significant recommendation for planning the future of the Institute is that, as soon as adequate resources are available, it should sponsor multidisciplinary networks of individuals whose research interests involve a common theme and whose efforts will be mutually supportive. The conclusion that the Institute should begin operations in the network mode is based on the following considerations:

1. Networks comprised of the most productive individuals and appropriately qualified students and offering workshops, strengthened communications, and a central campus staffed with non-resident, visiting, and permanent faculty can begin immediately to meet an increasingly urgent need to better organize and nurture interdisciplinary efforts at the forefront of research.
2. Such a program will offer prompt benefits not only to the participants and their research programs but also to their home institutions.
3. The sponsorship of networks on selected themes will serve an important purpose by better defining and emphasizing the importance of various emerging syntheses which tend to be fragmented and overlooked within the conventional disciplines.
4. The Institute will benefit from interactions with network participants in the careful identification and recruiting of senior faculty, junior staff, and graduate students.
5. The Institute can deliberately explore the relative merits of major themes suitable for long-term pursuit on the Santa Fe campus.

Accordingly, the Institute will devote its early resources to the formation of a few such networks each year while continuing to move toward full-scale operation as a teaching and research Institute. Even after reaching its full growth, the Institute will probably continue to maintain and expand such networks as a necessary means to strengthen vital parts of the scientific enterprise.

Some of the network themes proposed for early consideration include a program on theoretical neurophysics; the modeling of evolution, including the evolution of behavior; strategies to model troublesome states of minds and associated higher brain functions; nonlinear systems dynamics, pattern recognition, and human thought; fundamental physics, astronomy, and mathematics; archaeology, archaeometry, and forces leading to extinction of flourishing cultures; an integrated approach to information science; and the heterogeneity of genetic inventories of individuals.

Looking to the longer term, the Institute will plan to develop a campus which is large enough to provide sites for nearby, independent academic organizations representing social, political, and behavioral sciences and parts of the humanities. As experimental, computational, and mathematical tools grow in capacity, it is possible to envision a time, not far off, when the rigor of the hard sciences and elements of human experience and wisdom will be joined more effectively together so that we can better model and hope to understand the most complex and interactive systems of all, those which govern our bodies and brains and those developed within past and present societies which shape and govern much of our lives. There can be little doubt that the upsurge in sophisticated experimentation, the explosive growth of very large scale, parallel processor computing, the increasing capability of models which treat non-linear dynamic processes, and the development of new machine languages and algorithms far exceeding the power of those now in use will generate enormous forces for achieving noble or destructive ends. They must be wisely directed. With wisdom, the diffusion of the hard sciences into what are now considered the soft sciences may well become the most important achievement of the twenty-first century. To help insure that the next generations can rise to this challenge, the Institute must strive to promote a unity of knowledge and a recognition of shared responsibility that will stand in sharp contrast to the present growing polarization of intellectual cultures perceived so well by C. P. Snow nearly a generation ago.

It was not part of the workshop's agenda to consider the problem of obtaining the financial resources required to realize its plans. Some concern was voiced that the widespread demand for increased support throughout academia would make it extremely difficult to establish any new enterprise. However, a consensus was evident that the need and anticipated benefit easily justified the projected cost, that the time is ripe, and that these new ideas should be put forward without delay. Encouraged by these views the Institute has established a full-time development office and will actively pursue a fund-raising campaign until adequate resources are obtained.